ADMIRABLE EVASIONS

Theodore Dalrymple

ADMIRABLE

EVASIONS

How Psychology Undermines Morality

ENCOUNTER BOOKS · *New York ·London*

© 2015 by Theodore Dalrymple

First American edition published in 2015 by Encounter Books, an activity of Encounter for Culture and Education, Inc., a nonprofit, tax-exempt corporation.
Encounter Books website address: www.encounterbooks.com

Manufactured in the United States and printed on acid-free paper. The paper used in this publication meets the minimum requirements of ANSI/NISO Z39.48—1992 (R 1997) (*Permanence of Paper*).

FIRST AMERICAN EDITION

LIBRARY OF CONGRESS CATALOGING-IN-PUBLICATION DATA

ISBN 978-1-64177-188-7 (paperback)

Dalrymple, Theodore.
Admirable evasions : how psychology undermines morality / by Theodore Dalrymple.
    pages   cm
Includes bibliographical references and index.
ISBN 978-1-59403-787-0 (hardcover : alk. paper)
ISBN 978-1-59403-788-7 (ebook)
1. Self-perception. 2. Self-reliance. 3. Psychotherapy. 4. Psychoanalysis.
5. Psychology—Moral and ethical aspects. I. Title.
BF697.5.S43D35 2015
150—dc23
2014037247

CONTENTS

FOR RICHARD LATCHAM

It is a poor centre of a man's actions, himself.

FRANCIS BACON

To be beautiful means to be yourself. You don't need to be accepted by others. You need to accept yourself.

THÍCH NHẤT HẠNH

... an admirable evasion of whoremaster man, to lay his goatish disposition to the charge of a star!

WILLIAM SHAKESPEARE

PREFACE

In the misfortunes of our friends, wrote the duc de La Rochefoucauld nearly three and a half centuries ago, there is something not entirely unpleasing. When we read this for the first time we experience both a shock and a sense of recognition. Something discreditable about us has been put into words that, if we had reflected a little harder or more honestly upon ourselves, we should have known all along: and henceforth we shall never be able to pretend that we are other than complex and contradictory beings.

La Rochefoucauld was able to put into words what anyone could have known by "attending to the motions of his own mind," as Doctor Johnson put it. Have human beings progressed beyond this in their self-understanding? It is my contention that they have not, and that our vaunted or pretended progress, amazing though it may be from a technological point of view, is actually a retrogression in honesty and sophistication. Psychology is not a key to self-understanding, but a cultural barrier to such under-standing as we can achieve; but it is my belief that we shall never be able entirely to pluck out the heart of our mystery. Of this I am glad rather than sorry.

CHAPTER ONE

If all the antidepressants and anxiolytics in the world were thrown into the sea, as Oliver Wendell Holmes Sr. once suggested should be done with the whole of the pharmacopoeia, if all textbooks of psychology were withdrawn and pulped, if all psychologists ceased to practice, if all university departments of psychology were closed down, if all psychological research were abandoned, if all psychological terms were excised from everyday speech, would Mankind be the loser or the gainer, the wiser or the more foolish? Would his self-understanding be any the less? Would his life be any the worse?

It is not, of course, possible to give a definitive answer to these questions: the experiment cannot be done. But it would be a bold man who claimed that Man's self-understanding is now greater than that of Montaigne or Shakespeare. How many of us would dare to claim in public that he had greater insight into his fellow creatures than the Swan of Avon? He would be laughed down immediately, ridiculed and ignominiously driven from the platform: and quite rightly so. Such arrogance would have its reward. As to life having improved, how much of the improvement is attributable to psychology? We owe incomparably more to improved sewers than to psychology.

Yet implicit claims to superior knowledge and understanding are by no means uncommon. More than one school of psychology has claimed to have achieved deeper insight into human nature, conduct, emotion, and distress than ever before. In 1802, the French philosophical physiologist Pierre-Jean-George Cabanis said confidently that the brain secretes thought as the liver secretes bile. Two hundred years later, the acclaimed neuroscientist V. S. Ramachandran said essentially the same thing, though in more words, as if verbosity indicated progress:

Even though it is common knowledge, it never ceases to amaze me that all the richness of our mental life – all our feelings, our emotions, our thoughts, our ambitions, our love lives, our religious sentiments and even what each of us regards as his or her own intimate private self – is simply the activity of these little specks of jelly in our heads, in our brains. There is nothing else.

So everything in human self-understanding is over bar the shouting: only the details have yet to be filled in. Before long, if there is sufficient research funding, there will be no more puzzles and no unpleasant surprises, no agonizing dilemmas in human existence; the question of the good life will have been settled once and for all, indubitably and scientifically, without the necessity of endless and unprovable metaphysical speculations. To understand all will no longer be to forgive all, for there will be nothing to forgive; everyone will behave reasonably in the first place, which is to say, in accordance with the dictates of the scientifically proven good life. History will come to an end, this time not by virtue of the triumph of liberal democracy throughout the world, but by that of the triumph of psychology and neuroscience. Man will no longer pass on misery to Man, as in Larkin's poem; he will pass on knowledge instead, knowledge and wisdom being of course by that stage coterminous. Indeed, knowledge will secrete wisdom as the liver secretes bile.

I don't believe it, and I'm not sure that I would want to live in such a world if it were true. How dull everything would be! Life would be a perpetual Caribbean cruise aboard a luxury liner on a calm sea in clement weather. Mankind would be bored for lack of causes of unhappiness and would soon sink the boat on which he was cruising: for Man is not so much a problem-solving animal as a problem-creating one. Pascal said that all of Man's misfortunes come from his inability to sit quietly in a room: but he did not claim to have found a way to enable him to do so, or suggest that this inability comes from anything other than his inherent nature.

The first psychological scheme of the twentieth century to provide the common man with the illusion of much expanded, if not yet complete, self-understanding, together with the hope of an existence free of inner and outer conflict, was psychoanalysis. Then came behaviorism, after which there was cybernetics. Sociobiology and evolutionary psychology were next; and now neuroscientific imaging, together with a little light neurochemistry, persuades us that we are about to pluck out the heart of our mystery. Suffice it to say, by way of deflation of exaggerated hopes and expectations, that 10 percent or more of the population now takes antidepressants, a figure all the more remarkable as the evidence is lacking that they, the antidepressants, work except in a very small minority of cases; rather the reverse. That they are taken in such large quantities is evidence more of dissatisfaction with life than of increased understanding of its causes, as well as of the spread of superstition regarding neurotransmitters and so-called "chemical imbalances." These latter are to the modern person what alien spirits to be exorcised or the ego, id, and superego once were: things not seen but strongly believed in, as providing explanations for unwanted feelings, experiences, and behavior, as well as the hope of their elimination. Superstition springs as eternal in the human breast as hope.

So absurd does Freudianism now seem to us, so self-evidently false, that we forget what a hold it had on our self-conception only a few decades ago. W. H. Auden was right in saying, in his poem to mark Freud's death, that he was "a whole climate of opinion," and that if he was, in the opinion of the poet, "often . . . wrong and, at times, absurd," he was nevertheless working along the right lines and had much extended our self-understanding, for:

> in a world he changed
> simply by looking back with no false regrets;
> all he did was to remember
> like the old and be honest like children.

It would be difficult to put in a few words anything more inapposite about Freud, anything in fact more opposite to the truth than these lines, which capture so precisely and at the same time endorse the illusion of an age.

Freud was undoubtedly brilliant, a good writer and a very cultivated man, but his career, certainly once he stopped looking into the nervous system of eels, belonged more to the history of techniques of self-advancement and the foundation of religious sects than to that of science. It is historically certain that he was a habitual liar who falsified evidence in that way that Henry Ford made cars; he was a plagiarist who not only did not acknowledge, but actively denied, the sources of his ideas; he was credulous of evident absurdities, as his relations with Wilhelm Fliess prove; he was a self-aggrandizing mythologist and a shameless manipulator of people; he could be financially grasping and unscrupulous; he was the founder of a doctrinaire sect and a searcher-out and avenger of heresy who would brook no opposition or competition, and who called down anathema on infidels as intolerantly as Mohammed; in short, he was to human self-understanding what Piltdown Man was to physical anthropology. Insofar as Freud was sensible or profound, when for example he said that the maintenance of civilization depended upon restraint and the deliberate frustration of raw desire, no deep analysis of the human psyche was necessary to reach such conclusions, for they were available to any reasonably intelligent person who took the trouble to reflect for a moment upon the human condition; nor were they original, very far from it; they were the commonplaces of a million sermons.

Freud's claims to have been a scientist do not stand up to scrutiny for a moment, and his writings are now so unconvincing that it is a historical conundrum as to how anyone could ever have been convinced by them or to have taken them seriously in the first place. (When he arrived in England as a refugee from Austria he was immediately made a Fellow of the Royal Society, the highest scientific honor the country had to offer.) To read a prolonged case history by Freud is to wonder at the non sequiturs, the leaps of faith, the illogic, the arguments from authority in which they abound, but which were not, apparently, apparent to generations of readers. And although Freud was personally conservative in his manner and morality, except where his incestuous adultery with his sister-in-law was concerned, his effect, if not his intention, was to loosen Man's sense of responsibility for his own actions, freedom from responsibility being the most highly valued freedom of all, albeit one that is metaphysically impossible to achieve. For Freud powerfully alienated men from their own consciousness by claiming that what went on in their conscious minds was but a shadow play, and that the real action lay deep beneath it, all undiscovered (and undiscoverable) without many hours of talking about oneself in the presence of an analyst who might from time to time offer an interpretation of the real meaning of nonacceptance, which would itself be interpreted as resistance in need of further analysis, and so on more or less *ad infinitum*.

It had long been no secret that men do not always act for the reasons that they say they do, that they easily deceive themselves as much as others, that their motives are frequently (though not always) secret, mixed, and discreditable, that they project onto others the very illicit wishes and desires that they themselves have. King Lear, whose words were written nearly three hundred years before Freud's "revelations," said:

Why dost thou lash that whore? Strip thine own back.
Thou hotly lust'st to use her in that kind
For which thou whipp'st her.

No one, surely, can have reached adulthood without having realized that human existence is not an open book, that much about ourselves and others remains hidden from us; at the same time, it should not have escaped creatures endowed with reason and powers of self-reflection that it is often possible by means of taking thought to discover much about ourselves and others that was not immediately evident to us, and that the better and more honest we are at taking thought, the more we shall discover.

But psychoanalysis is not so much reflection as a kind of shallow gnostic divination. It starts off with the hypothesis that all thoughts are born equal, at least in deeper psychological significance, and that the conscious attempt to discipline them, to winnow the true from the false, the important from the trivial, the useful from the useless, actually inhibits or prevents the achievement of self-knowledge. The only discipline that is necessary for the achievement of the latter is the abandonment of discipline (free association), which admittedly is not easy to achieve for intelligent people who have hitherto acted on the supposition that disciplined thought is desirable and important.

The result was and is all too predictable. Psychoanalysis, as well as death, becomes a bourn from which no traveler returns: and like anything indulged in for a long time, concern over the small change of life becomes a habit, and an irritating one, that inhibits interest and taking part in the wider world. It is a poor center of a man's attention, *himself*; compared with psychoanalysis, haruspicy or hepatoscopy (divination by entrails or the liver of sacrificed animals) is harmless to the character, for though it is absurd, it at least is limited in time. Psychoanalysis becomes an ingrained habit of mind that must itself be overcome, often with the greatest difficulty, if the person undergoing it is not to torment himself and others for the rest of his life with seeking the hidden meanings of utterances such as "Good morning" and "How are you today?" (so easily interpreted as a wish that the person thus addressed may drop dead). True, Freud once said that sometimes a cigar is only a cigar – not a coincidental choice of object to remain only itself, for he was an inveterate smoker of cigars – but he did not provide a criterion for discerning when a cigar is only a cigar and when it is a phallic symbol undergoing fellatio (presumably when it was smoked by others, not by himself). It is hardly surprising that the world comes to seem for the analysand an infinite regress of symbols, a labyrinth, a hall of mirrors in which images of himself stretch into the confined infinitude of his mirrored chamber. If psychoanalysis had been invented by cavemen, Mankind would still be living in caves.

I do not mean by this to imply that the human mind is straightforward, that we are aware immediately and at all times of all the reasons for our thoughts and actions. A moment's reflection should be enough to show that this cannot possibly be so. We do not even know where our thoughts come from: but we know that we can think, and that we can direct our thought and discipline thoughts once they have arisen, check them for veracity, decency, consistency, etc.

It is also true that our utterances, even trivial ones, can sometimes unintentionally reveal something important about the way we think. For example (an example I have used before), people who stab someone to death with a knife often, indeed usually, say "The knife went in." It is hardly a wild surmise that this way of putting it distances the perpetrator from his responsibility for his action, a disagreeable thing to reflect upon and full of the direst legal consequences, turning a voluntary and even long premeditated action into a chance event determined by the disposition of physical objects. The knife guided the hand rather than the hand the knife; as Edmund (in *King Lear*) puts it, "An admirable evasion of whoremaster man, to lay his goatish disposition to the charge of a star!" We all do it at times in our lives; indeed our first response when accused (by ourselves or others) of wrongdoing is to find exculpatory circumstances to explain it, or rather to explain it away. But thanks to the mind's marvelous and subtle ability to think in parallel tracks at the same time, we have a still small voice telling us that our excuses are bunk. That is why so much anger is both real and simulated, spontaneous and deliberately generated, at the same time. But we don't need psychoanalysis to show us any of this; and it is equally obvious that we must exercise judgment in attributing unacknowledged motives. Sometimes an extenuating circumstance is just an extenuating circumstance (no one, of course, seeks to extenuate his good deeds).

A doctrine or philosophy insinuates itself into a culture by means of rumor as much as by persuasion occasioned by reading its founding, or even subsequent, texts: that is why Auden's "climate of opinion" is so accurate. And the lessons drawn from the doctrine thus insinuated may not be such as the founder meant or would endorse. In the case of psychoanalysis, the lessons popularly drawn were that the slightest utterance was of the deepest significance and everything said was instinct with hidden meaning; that all human desires were ultimately of a sexual nature; that human desires acted in a hydraulic fashion, and like liquid could not be compressed, so that if they were not fulfilled they would make themselves manifest in some other, pathological way, hence frustration of desire was both futile and dangerous; and that, once the supposed biographical cause of a pathological symptom, buried deep in the individual's past, was revealed after much digging around in the dregs of the mind, it would cease by itself, without any need for the individual's effort at self-control.

The first lesson, the deep significance of every utterance and its supposed suffusion with hidden meaning, naturally conduces to a combination of triviality and paranoia: triviality because it dissolves the very distinction between the trivial and the significant, the former coming much more easily to the voice than the latter; and paranoia because hidden meanings are sought everywhere since they are presumed to exist and to be in need of interpretation. Neither good nor evil acts are taken at face value any longer, but are assumed to be *really* other than they appear, usually their opposite in fact. Thus kindness (in others) becomes hidden aggression and rudeness (in oneself) becomes a defense against the overwhelming strength of one's own generosity of feeling. Triviality has, of course, been given a tremendous fillip by the so-called social media, in which the social contract has been rewritten to read "I will pretend to be interested in your trivia if you pretend to be interested in mine" – which of course I don't really believe to be trivia in the first place, at least not *my* trivia. I am a man, wrote Terence; I find nothing human uninteresting. Thanks to the progress wrought in human self-understanding by psychoanalysis, our dictum has changed. It is now: I am a man; I find nothing about me uninteresting.

That desire, if not fulfilled, will lead to pathology makes of self-indulgence man's highest goal. It is a kind of treason to the self, and possibly to others, to deny oneself anything. Thus there is a time and place for everything, that time being now and that place being here. It is hardly surprising that such an attitude should end with widespread and enormous personal indebtedness. When a new credit card was launched in Britain, its advertising slogan was that it would "take the waiting out of wanting." What is waiting for what is wanted but a form of frustration? And if frustration of desire is the root of pathology, then it follows that the credit card must be the cure of much pathology. Did not Blake say "He who desires but acts not, breeds pestilence" and "Sooner strangle an infant in its cradle than nurse unacted desires"? The road to heaven is paved with fulfilled desires, and to hell with frustrated ones. How terrible then for the parents of children to stay together just for the sake of duty when one of them "needs his space" because "it just isn't working." As one patient of mine put it soon after he had strangled his girlfriend, "I had to kill her, doctor, or I don't know what I would have done." Something serious, perhaps.

As to the automatically curative nature of psychological buried treasure, belief in it is now almost as widespread as belief in miracle-working images once was among the religious. In fact, it is a sovereign excuse for continuing to do what you know you should not do, for it is obvious that the supposedly liberating buried treasure can remain buried forever, however long you dig. The fact that your bad behavior or habit, whatever it is, continues despite psychological excavation is *ipso facto* evidence that the buried treasure has not been found and that the search must go on because it is buried deeper. A ludicrous and dishonest *pas de deux* then takes place in which the therapist and the patient search for what is not there, and since absence can never be proved, it is hardly surprising that Freud wrote a paper toward the end of his life, "Analysis Terminable and Interminable," that raised the possibility, no doubt alluring to some patients, of talking about yourself forever.

It might of course be said that the popular deformation of an idea or practice does invalidate that idea or practice; but I am here concerned, in this little book, precisely with the *overall* effect in society of psychology as a discipline or way of thought. In any case, when we look at the effects of psychoanalysis on those who may be presumed to have had a more detailed, true, or accurate knowledge of it (if, that is, knowledge of any doctrine as slippery as that of psychoanalysis can be called true or accurate), the scene is not more encouraging. For example, nine of the first Viennese psychoanalysts, one in seventeen of them, committed suicide. The personal relations of these first psychoanalysts, moreover, were not such as would recommend themselves to anyone, consisting as they did largely of backbiting, betrayal, envy, denunciation to the authorities (i.e., to Freud), excommunication, and cuckolding. Such relations would be of no account if the subject matter of the group had been meteorology, shall we say, or astrophysics, but we are surely entitled to expect that people who set themselves up as unprecedentedly expert in human relations, as having special insight into the psychology of their fellow beings, will have shown special wisdom in the conduct of their own lives. Of course, it is possible that they were peculiarly unsettled people in the first place, which explains why they were drawn to psychoanalysis; but the very least that can be said is that contact with psychoanalysis did not seem to have effected much improvement. Nor does psychoanalysis seem to have conferred wisdom or insight into wider matters: Freud was claiming as late as 1938 that his *real* enemy, the enemy he was really afraid of, was not the Nazis but the Catholic Church.

As for analysands, you meet some who claim that their lives were much improved by their analysis, but this is as little evidence of the truth or value of psychoanalysis as is the recent convert to Islam's opinion of the Koran as proof of the prophethood of Mohammed. A little bit of what you fancy may do you good, but it doesn't make it true.

In my experience, at least, analysis has at least as many harmful as good effects (I put it mildly). It turns people permanently inward and guts their language of individuality, replacing particularity by a kind of impersonal jargon. They often appear to have undergone a strange type of brainwashing. The philosopher Karl Popper accused Wittgenstein of always polishing his glasses but never looking through them; the same might be said of people who have undergone psychoanalysis. And

Freud hit upon a brilliant method of keeping analysands loyal (I don't mean that he did this deliberately): if you make a treatment long and expensive enough, people will always find that it did them at least some good, for otherwise they would have wasted their time and money, and would look foolish – even to themselves.

Recently I was sent for review a book by a woman who had been in analysis for twenty years, with four or five sessions a week, in all about four thousand. Four thousand hours of talking about oneself! Full marks for endurance, if not for choice of subject matter. Whether it did her any good is, of course, a question that cannot be answered definitively. What she would have been like without it must be a matter of fruitless speculation. The author, Barbara Taylor, is a historian who suffered no serious traumas in her life except that of her own personality and the consequences thereof. In the book she recorded some of the interchanges she had with her analyst, who seems to have been a lot more communicative if not necessarily more profound in his utterances than most orthodox analysts. I presume she must have written them down immediately after they took place:

SHE: Why [do I keep coming to the analyst]? Is it just to torture myself?
HE: Sometimes.
SHE: Why else? Why else would I keep coming here? I'm just making myself suffer more! Why else would I keep coming here?
HE: You have different reasons at different times.
SHE: What reasons?

HE: Well … to get revenge on your parents. To get revenge on me, their current representative.

SHE: Ah. Yes. [I have heard this so many times, it doesn't mean anything.]

HE: And because you're waiting for a miracle.

SHE: A miracle? [This is more interesting. Is he offering? Maybe there's something he can do for me that he hasn't done yet.]

HE: Yes, the miracle that will make you the baby your mother really wanted, the sort of baby she could really love, so that she would look after you properly.

SHE: Oh. That miracle. [Why is there never anything new?]

HE: And sometimes it's because you want to know the truth.

SHE: The truth? What truth?

HE: About what happened to you.

SHE: I know what happened to me.

HE Do you?

SHE: [Do I?]

This, I presume, must have been one of the highlights of twenty years of analysis, or it would not have been selected for inclusion in the book. Behind the self-obsession of the analysand and the portentous banality of the analyst's interjections lies the idea, self-exculpatory, that we are victims of our past, about which we can do nothing (unless, that is, we pay an analyst for four thousand sessions). Indeed, a British psychoanalyst called Adam Phillips wrote in a recent book, *Becoming Freud*, that childhood is inherently catastrophic and the past is, in his typically inelegant phrase, "unrecoverable from." (Psychoanalysis, it seems, does wonders for a man's prose style: it renders it labyrinthine without subtlety.) There is no place, then, for human agency, except the kind that leads you to talk about yourself in the presence of another for twenty years. Shallowness can go no deeper.

CHAPTER TWO

Though for several decades Freudianism ruled the psychological roost, and won the allegiance of those – always a large number – who seek to escape from the terrible burden, though also the glory, of being human (that is to say, of having to choose at all times how to act and respond to circumstances, and therefore of being at least partially responsible for their fate), it was never quite without opponents, enemies, or resistance. (I do not here mean resistance in the psychoanalytic sense, the refusal by an analysand to accept an interpretation by his analyst, a refusal taken by the latter as confirmation of its truth: not a manner of proceeding conducive to self-criticism among analysts, to put it mildly.)

No, resistance to psychoanalysis as a doctrine and a method arose because of its intellectual inadequacies, which were, or should have been, evident from the first. It provided no criterion of truth to distinguish between a true and a false interpretation, or even between a plausible and an implausible one, itself a deficiency so serious that it vitiated the repeated claims of psychoanalysis to be a science. Interpretations were based upon the theory and the theory (allegedly) upon interpretations, a circularity from which there was no escape: for despite Malinowski's expedition to the Trobriand Islands to find the Oedipus complex among their Stone Age inhabitants, there never was and never could be any independent evidence of the intellectual constructs of psychoanalysis. It was a closed system in which one had a priori to believe as a kind of act of faith.

In reaction to the intellectual indiscipline and clinical impotence of psychoanalysis, a mindless psychology – a psychology that excluded mind – became quite popular, eventually developing into an orthodoxy, at least among psychologists, with a little church of its own. The data of consciousness, pronounced the popes of behaviorism, were not susceptible to scientific verification; therefore they should be excluded from scientific enquiry. Instead, psychology should study only verifiable and measurable inputs and outputs, stimuli and responses, for whatever happened between input and output, stimuli and responses, was inaccessible to verification and measurement. It was worse than the black box of Flight MH370: it was not merely unfound, it was inherently unfindable.

What started as methodology became ontology. It is an old adage of medical diagnosis that absence of evidence is not always evidence of absence, but the behaviorists ignored this sage call to modesty. Instead, they began to believe that stimulus and response were all there was to human life, that everything human could be explained by it. Again, this should have been laughable, but it was taken by many with the utmost seriousness. An intellectual could almost be defined as a person who follows an argument to an absurd conclusion, and believes the conclusion.

Behaviorism was not without its successes, however,

and not just the institutional ones that psychoanalysis undoubtedly enjoyed. Its theory was used, for example, to teach pigeons to play table tennis. Who would have thought you could teach pigeons to play table tennis? Actually, I am not sure that you can teach it: for a game of table tennis involves more than merely batting a ball back and forth over a little net stretched across a green table, astonishing though the ability to do even this is in the case of pigeons. Among other things, a real game of table tennis entails the desire to win something as abstruse as a game, a desire which it is difficult to believe that pigeons can have; and also to know the rules. A pigeon is unlikely to keep score or celebrate victory when it is the first to reach twenty-one points. In other words, it will show no behavioral sign of having understood the *meaning* of what it is doing. Behaviorism entails the systematic denial of meaning, a denial which does violence to both the evidence and the everyday experience of humanity.

It is perhaps hard now to believe, but behaviorists claimed, and induced many people to believe, that the whole of human behavior could be explained by the scheme of stimulus and response, leading to aversion or reinforcement. Their assumptions and generalizations now appear to us naïve to the point of being ludicrous. In what might be called the Behaviorist Manifesto, written in 1913 by the founder of behaviorism, John B. Watson, we read:

The position is taken here that the behavior of man and the behavior of animals must be considered on the same plane; as being equally essential to a general understanding of behavior. It can dispense with consciousness in a psychological sense.

And while he admitted that:

Psychology as behavior will, after all, have to neglect but few of the really essential problems with which psychology as an introspective science now concerns itself

he went on optimistically (or was it pessimistically?) to state that:

In all probability even this residue of problems may be phrased in such a way that refined methods in behavior (which certainly must come) will lead to their solution.

Watson was nothing if not a great jumper to conclusions, as great in his own way as Freud in his. In his famous 1920 paper on the case of Little Albert, he and his coauthor, later his wife, described how he conditioned an eleven-month-old baby to become fearful of a white rat by making a loud clanging noise on presenting the rat to the baby. He tested the baby a month or so after the conditioning had ceased, but still the baby showed a conditioned response to the rat, albeit muted by comparison with the conditioned response immediately after he had been conditioned. Watson says confidently:

Emotional disturbances in adults cannot be traced back to sex alone. They must be retraced along at least three collateral lines – to conditioned and transferred responses set up in infancy and early youth in all three of the fundamental human emotions.

Humans, in the view of the behaviorists, are but glorified pigeons playing table tennis. When someone's behavior was explained by the mental processes within him, one of the most influential of the subsequent behaviorists, B. F. Skinner, asked why anyone should seek the explanation of the explanation. According to him, all talk of subjective experience and consciousness stifled further enquiry. This, of course, is mistaken. Psychoanalysis might have made outrageous claims for itself, premature announced understanding, but no one could say that it did not seek to explain the thoughts that people had. Skinner's remark, moreover, suggests that he thought he had found, if not the complete explanation of human life, at least the fundamental principle of such an explanation. All that remained to be filled in was the detail: for example, how Beethoven's late quartets were a conditioned response to Beethoven's then circumstances.

Skinner's remark also suggests that an explanation of the type he sought actually exists and is available to human beings. This might not be so for metaphysical reasons. If the metaphysical arguments that lead to this conclusion were correct – that in effect, mankind will never be able to pluck out the heart of its own mystery as if men were laboratory specimens – then efforts to explain mankind to itself, of the kind of which Skinner might approve, are fundamentally misguided and destined to fail. Unfortunately, such error would not be confined to the intellectual sphere, but would be likely to spill over into the "real" world of action and policy. Skinner himself was quite clear about this: he thought that a society could be erected, if not now then at some time in the not very distant future, on behaviorist principles. All we had to do was choose a goal and condition people to achieve it. There were, of course, no metaphysical questions as to which goal to choose, or who was to condition the conditioners. Skinner wrote a utopian novel, *Walden 2*, to illustrate his theory of a rationally designed society. It seems to me about as realistic as the socialist utopias started in late nineteenth-century Paraguay. Perhaps Skinner's greatest achievement was in stimulating Anthony Burgess's response to behaviorism in *A Clockwork Orange*, in which Alex is conditioned out of his love for classical music by means of electric shocks administered while he hears Beethoven's Ninth Symphony.

Black box psychology now seems to us almost as bizarre a cultural product as phrenology or spiritualism. Its ferocious determination to eliminate from study the one thing that (as far as we know, though somewhere in the universe there may be other beings like us) gives worth to the universe, namely human self-consciousness, outpuritans the Puritans. From the claim that the contents of consciousness cannot be studied scientifically, and therefore should not be the object of study at all, it was but a short step to denial of either the existence or the importance of human self-consciousness. Of course, if it were really true that the contents of consciousness could not be studied scientifically, another possible conclusion might be drawn: that the attempt at a scientific psychology that explains Man to himself is doomed to failure. The very thing that is most important to us is the very thing that is out of our reach. The history of psychology bears some resemblance to the myth of Sisyphus: punished for his deceitfulness, he was made to roll a rock up a hill, only for it to roll back to the bottom just as he reached the brow of the hill with it. Psychology repeatedly announces huge advances in human self-understanding, only for the announcement to be shown to be premature, whereupon another school promptly steps into what might be called the self-promotional breach.

This is not to say that behaviorism was in no respects an improvement on Freudianism. It had its limited successes. Though the administration of aversive stimuli to homosexuals and alcoholics did not "cure" them (electric shocks on the presentation of sexually arousing images to the first, and of a drug, apomorphine, that caused nausea on the presentation of alcohol to the latter, on the grounds that a little bit of what revolts you does you good), specific phobias (such as those to spiders) *did* respond relatively well to treatment on behaviorist principles. This meant that patients no longer had to seek, tediously and expensively, for the psychological buried treasure of the symbolic meaning of their fear of spiders, but could actually be helped to lose their irrational fear, which on occasion could be crippling. Confronted with evidence that this was the case, the psychoanalysts promptly argued, with the inventive cunning that each of us possesses when a pet theory of ours is refuted by the evidence, that unless the buried treasure were found, the phobia would be replaced by some other symptom, possibly worse than the original one. No such symptom substitution has ever been found, but of course for psychoanalysis some symptoms lie too deep for visibility, except to itself.

Behaviorism was a symptom of Man's perpetual metaphysical impatience. Each man has only a lifetime in which to understand himself, and as with many foreign expeditionary wars, there is a strong temptation to declare victory and go home. From the fact that conditioning undoubtedly exists the somewhat broad conclusion is drawn that *only* conditioning exists, and that it is the explanation of all things human. The sheer preposterousness of this never really struck its proponents, who ruled for a time in the psychology departments of some of the best universities; nor did its violence to common human experience. They treated human life as if it were a vastly expanded case of arachnophobia. Life was henceforth to be made perfect by a judicious combination of electric shocks and food pellets.

Behaviorism was but one instance of a terrible temptation for all intellectuals, namely that of nothing-but-ism. History is nothing but the clash of class interests, human behavior is nothing but a response to economic incentives, etc., etc. Of course, it doesn't require much knowledge or reflection (or for that matter self-examination) to agree that people often respond to economic incentives, but as an explanation of *everything*, and therefore of all history, it is as preposterous as the belief that the Mass in B Minor is really only a sublimation of Bach's unacknowledged sexual desire for his mother, or a conditioned response to the death of Augustus II, Elector of Saxony.

The search for the elementary particles of human existence does not so much obey the biblical injunction *seek, and ye shall find* as follow the less glorious procrustean epistemological principle *find, and ye shall seek*. Once a metaphysically impatient person has found his guiding explanatory principle, he manages to explain everything by it: and the more he explains, the more intoxicatingly explanatory his principle appears to him.

Another curious phenomenon in the history of psychological thought is the recurrence of the powerful personality or egomaniac who denies the reality of personal identity. David Hume, who was definitely not an egomaniac, denied the reality of personal identity because he could find no I that was independent of the current contents of that I's consciousness. What exactly is that I that connects me of today with me of yesterday? It cannot be conceived of, says Hume, in the absence of all experience whatsoever; therefore the I is nothing subsisting. You cannot step into the same river twice, said Heraclitus; you cannot be the same person at two different moments. Indeed, if time and experience in that time are indivisible, you cannot be a person at all (you cannot step into the same river even once, let alone twice). One wonders whether someday a criminal, or a lawyer arguing on his behalf, will claim that the object in the dock – for to be consistent, his defender cannot call him a person – cannot be punished because he is not the person who did the deed. Indeed, there was no deed or act to be punished, because there was no beginning and no end to it, only a continuous flow, of which the division into portions, known as events, is inevitably arbitrary.

Where Hume wrote with irony, others write with earnestness. Metaphysics, said the late nineteenth-century idealist philosopher Bradley, is the finding of bad reasons for what we believe on instinct; but metaphysics has changed in the meantime, and is now the finding of bad reasons for what we cannot possibly believe however hard we try. All I can say is that the disbelief in the reality of consciousness or personal identity has never prevented anyone from copyrighting his book in which that unreality is argued; and I very much doubt that any author of such a book has ever been completely indifferent as to the bank account into which its royalties were paid.

Yet another characteristic of psychological thought is its overestimation of the light that pathology, both physical and psychological, throws on normal functioning, and thus on human existence as a whole. Freud, as is well known, began with hysteria and ended with civilization. Watson started with Little Albert and ended with the whole of human life. Physical pathology, by contrast, has a much more distinguished history. The brains of lunatics had long been examined for the source of their lunacy without much success, the assumption being that the difference between them and the brains of non-lunatics would reveal the *fons et origo* of madness and therefore, indirectly, of sanity; and eventually progress was made. The deduction by Fournier that general paralysis of the insane (GPI), responsible for 15 percent of asylum admissions in Britain and France in the second half of the nineteenth century (and an even greater proportion of the deaths in them), was a late complication of syphilis was a triumph of medical observation and deduction, and it was made even before the causative organism of syphilis was discovered and the neuropathology of late syphilis elucidated. (Incidentally, I strongly suspect that Ibsen kept abreast of Fournier's work, and used his findings in the plot of his play *Ghosts*.) Later, the treatment of syphilis became so successful that I have only once in my career seen a case of GPI, and he was a vicar of the Church of England.

Perhaps the most important single event in the history

of neuropathology was the finding by Broca of the area of brain damage associated with aphasia after a stroke, known to this day as Broca's area. This set off a kind of gold rush to find areas of the brain that were supposedly the organs of the faculty that was damaged or destroyed when they were injured traumatically or by some other pathological process. If an area of the brain were damaged and the person who suffered that damage showed a particular deficit of function, it was assumed by researchers that the area of the brain damaged was the seat of that function, as the heart is the pump of the cardiovascular system or the kidney is the organ of excretion. Not only did this understandable viewpoint underestimate the capacity of neural tissue to regenerate or repair (it was the doctrine at the time that all damage to the brain was both physically and functionally irreparable), but it failed to distinguish between a necessary and a sufficient condition. It was as if, having picked the legs off a fly and observed that the fly no longer flew, an experimenter were to conclude that its legs were a fly's organs of flight. Researchers thought that if only neuropathology were refined enough, all human functions, all human conduct and thought, could and would be chased to its lair, as it were, and would thereby have been satisfactorily explained. But the metaphysical problems, it is now easy to see, would have remained, even if (as is not the case) every little part of the brain carried out a discrete function, all of which functions added together making up a whole human being.

The famous case of Phineas Gage supposedly "proved" the neurological basis of social behavior, and of morality itself. Gage was the foreman in the construction of a railway in Vermont in 1848 when an explosion drove an iron bar through his skull, in the process damaging the front part of his brain. Miraculously he survived, and indeed recovered his physical health. But he was a changed character; whereas before the accident he had been a fine, reliable, upstanding citizen, religiously observant, he became afterward something of a psychopath, irresponsible, selfish, drunken, impulsive, and incapable of forethought. This story, which appears in almost all textbooks of neuropsychology, and makes of Gage almost the Romulus and Remus of neuropsychology (as if he had volunteered for his injury so that science might benefit), is used to demonstrate the "seat" of the qualities that Gage lost after his accident.

That the case of Phineas Gage belongs at least as much to the history of metaphysical belief as to that of neuroscience is suggested by the eagerness with which it was taken up, despite the flimsiness of the historical evidence on which it was based. Detailed historical research published in 2000 by an Australian neuropsychologist, Malcolm Macmillan, showed that the by then traditional story of Gage's accident and its sequelae partook as much of mythology as of established fact, and indeed performed the function of the life of a saint in the justification of a religious faith more than a rational, evidential foundation of a scientific theory. The fact that the historical evidence available to Macmillan had been available more or less from the first demonstrates a *will* to believe on the part of those who overinterpreted the official story on such flimsy historical evidence, taken on faith, and becoming true by repetition; there is even a children's book recounting the "official" neuropsychological story of Phineas Gage, presumably with the intention of inducting children into the faith – the belief that, except for details, Man has already plucked out the heart of his mystery, and that no unsolved metaphysical problems remain. This is a point to which I shall return.

CHAPTER THREE

But here instead I return to behaviorism and its later offshoot, cognitive behavioral therapy (CBT). The absurdity of treating humans as glorified Pavlov's dogs having become evident and inescapable, a new element was then added to the recipe: *thought*. This startling innovation gave birth to CBT. Human travails, including counterproductive or destructive and repetitious patterns of conduct, were the result of mistaken and equally repetitious patterns of thought which, if interrupted, stopped, and changed, would alter behavior for the better. According to CBT, you could think yourself well. An unacknowledged progenitor of the technique was Émile Coué, the French pharmacist who realized the power of the placebo, and hence of autosuggestion. Interrupt gloomy thoughts by positive ones and you would soon feel much better: happiness is always but a mantra away.

"Cognitive-behavioral therapy," says a chapter of one textbook,[*] "seeks to improve functioning and emotional

---

[*] Christie Jackson, Kore Nissenson and Marylene Cloitre, "Cognitive-Behavioral Therapy," in *Treating Complex Traumatic Stress Disorders*, eds. Christine A. Courtois and Julian D. Ford.

well-being by identifying the beliefs, feelings and behaviors associated with psychological disturbance, and revising them through critical analysis and experiential exploration to be consistent with desired outcomes and positive life goals." If, as Buffon said, the style is the man himself, the style is also the technique itself.

For those with obsessions and compulsions or mildly depressive feelings, CBT works, and to that extent is a benefit to mankind. Its effects appear specific rather than general: other forms of treatment do not work so well, though of course it is impossible to perform double-blind trials of CBT without either patient or practitioner knowing what treatment he is having or giving. The Coué effect of belief and enthusiasm cannot be ruled out to account for the successes of CBT.

As is almost always the way with converts to anything, believers in CBT come wildly to overestimate its scope, and regard it as a panacea for most, if not all, the disgruntlements of Mankind. Nothing is good or bad, but positive or negative thoughts make it so. This soon translates into jobs for the boys (or as we must now say, persons): in Britain's centralized state health system, the National Health Service, there are now thousands of psychological wellness practitioners, "restructuring" the thoughts of their fellow citizens to relieve them of their mind-forg'd manacles and turn them into happy, productive citizens, needless to say with great claimed success. The new dawn has arrived.

Misery, however, has a tendency to rise to meet the means available for its alleviation, as Dr. Colin Brewer once pointed out. In other words, a supposed illness may increase in frequency as knowledge of its existence and the possibilities of its cure spread. And there are fashions in diagnosis as in everything else, even in physical diseases (some, such as floating kidney, have ceased altogether to exist, though the latter was supposedly cured by a hazardous operation, nephroplexy), let alone fashions in psychological disturbance. In the case of the latter, the situation is even more complicated because people can so easily talk themselves into disturbance. The humorist Jerome K. Jerome long ago pointed out, in *Three Men in a Boat*, that you can persuade yourself into the symptomatology of a hundred diseases merely by reading a medical textbook; how much easier is it to do so when the symptoms are explicitly in the mind! Indeed, it is part of the very foundation of CBT that this is what people commonly do. Therefore, the effectiveness of CBT is perfectly compatible with the spread in the population of the very condition for which it is supposedly the effective cure. Perhaps there is a prevalence of certain conditions, such as obsessive-compulsive disorder, that is "natural" and cannot be reduced, but one cannot help but wonder whether many conditions (for example, various kinds of eating disorders) spread in proportion as they are known about. La Rochefoucauld said that some people would never have fallen in love if they had not heard of it. It is possible, of course, that eating disorders once went completely unrecognized, which is why they seem so much more prevalent nowadays; but it is also possible that they really are more prevalent now, because of publicity. The Werther effect, after all, is well known: it was named for an epidemic of suicides by young men all round Europe after the publication of Goethe's *The Sorrows of Young Werther*, in which the romantic hero kills himself because of impossible love. It now refers to the increase in the number of suicides after publicity is given in newspapers or on television to a suicide. Research shows that the suicide of a celebrity has a Werther effect four or five times larger than that of an unknown person, and no better evidence of the shallowness of the roots of many tragedies could be offered.

Be that as it may, the Werther effect shows that imitation is potentially a cause of psychological disturbance, even serious disturbance, precisely because Man is a thinking reed, as Pascal put it (a weak and pliant thing, but nevertheless conscious and self-directed); it is therefore wrong in principle to treat the prevalence of psychological disturbance as a natural, raw fact, shall we say like the prevalence of hypothyroidism or Marfan syndrome; for the reflexiveness of Man's mentality means that factors such as imitation must be taken into account. No statement that a psychological disturbance has such-and-such a prevalence in such-and-such a population should be taken at face value, especially when it is a plea, as it so often is, explicit or implicit as the case may be, for more resources to treat it, the supposed prevalence having risen shockingly in the last few years. It is not merely that epidemiological searchers in this field can find what they are looking for; it is that they can actually *provoke* what they are looking for.

There are other ways to promote psychological fragility, of course; for example, by rewarding it. In countries with adversarial tort systems of civil law, plaintiffs have a vested interest in maximizing the harm that they have suffered from an accident of some kind. The problem with physical injury, from their point of view, is that recovery is a more or less objective process, difficult to deny once it has occurred. Psychological consequences of injury, on the other hand, are both easy to fake and difficult to disprove. The grosser forms of fraud may sometimes be discovered by the employment of private detectives, but anyone with access to the Internet and a reasonable degree of care and determination can act out any number of so-called neuroses. And, since the law's delay is long, and conduct that is assumed for long enough becomes part of the fabric of a person's character or personality, the person who fakes neurosis for long enough actually becomes neurotic. If you claim not to be able to concentrate or to leave the house, eventually your concentration will be destroyed or you will become housebound. Since most people do not like to think of themselves as frauds, the symptoms continue even after the case is settled: for if they, the symptoms, abated the moment a settlement of the claim were reached, their origins in fraud would stand revealed not only to others but to the plaintiffs themselves. The same is true, incidentally, of supposed physical symptoms after injury without evidence of physical pathology, which are therefore undisprovable by physical means; a wonderful exposition of the phenomenon is to be found in Andrew Malleson's book, not nearly as well known as it ought or deserves to be, *Whiplash and Other Useful Illnesses*. The real cause of much psychological disability, then, is the tort system, without which sufferers might be psychologically much more robust.

(This is not to claim that horrific events may not have severe psychological consequences for those who experience them, only to say that the apparatus of supposed recuperation and aftercare has profound effects upon the incidence of psychological consequences, often of a much less horrific nature. It may well be, then, that the overall effect of the apparatus is negative rather than positive, even though it is positive in some cases. Incidentally, the virtue of resilience or fortitude is the sworn enemy of that apparatus, which needs human vulnerability as a carnivore needs meat. When my wife had a miscarriage and refused the counseling that was offered her afterward, the person who offered it to her looked like an animal that had been cheated of its prey, and spoke to her as if she were being irresponsible, and would inevitably suffer the now self-inflicted consequences.)

CHAPTER FOUR

Paling beside the efforts of the tort system, however, are those of the social security system. There has long been a struggle to have psychological disorders treated by the system in the same way as physical illnesses, a struggle in which psychiatrists and psychologists have taken part without acknowledging that they (who see vested interests everywhere else) might themselves have a vested interest in the matter. The government agrees to it because this not only creates a new class of dependents upon itself, but allows it to minimize the rate of unemployment: for a sick person is not unemployed, he is sick. The easiest way to reduce unemployment, therefore, is by the stroke of doctors' pens.

In the United States a law has been passed, the Mental Health Parity and Addiction Equality Act, requiring insurance companies to be no more restrictive in their payments for psychological disturbances than in those for physical illnesses and injuries. This hardly matters to insurance companies that can pass on their costs to the insured, with a little bit extra just for profit; but it is very expensive for society as a whole. If there were only a slender chance of anyone suffering from such a disturbance, perhaps the parity act would not matter very much; but when you add all the annual prevalence rates given in the fifth edition of the *Diagnostic and Statistical Manual of Mental Disorders*, published by the American Psychiatric Association, it is clear that the average citizen suffers from at least two psychological disturbances per annum, and possibly quite a lot more. Assuming that half the population is psychologically healthy, whatever that psychological health may mean (my grandmother believed in weekly doses of castor oil, on the grounds that if you got rid of the bodily poisons, the mind poisons would look after themselves), the other half must suffer from at least four psychological disturbances annually. It is a wonder that any work is done in society other than the care of its incapacitated.

The expansion of psychiatric diagnoses leads paradoxically and simultaneously to overtreatment and undertreatment. The genuinely disturbed get short shrift: those with chronic schizophrenia, which seems most likely to me to be a genuine pathological malfunction of the brain, are left to molder in the doorways, streets, and stations of large cities, while untold millions have their fluctuating preoccupations attended to with the kind of attention that an overconcerned mother gives her spoiled child – with more or less the same results.

The notion that psychological disturbance should have parity with physical illness is likewise wrong in two directions, in that it both over- and underestimates it. General paralysis of the insane, dementia, or chronic schizophrenia are more terrible than an amputated limb, say (however undesirable or even tragic the latter may be), because they attack the very fundament of a person's humanity, his character and personality, his soul, if you like. Psychological illnesses undermine a person's capacity to reflect, think, and decide for himself; they render him, by degrees, precisely what overenthusiastic social engineers believe humans (other than themselves) to be: organisms without the capacity to decide for themselves. On the other hand, minor changes in equanimity are treated by patient and therapist alike with the utmost seriousness, as if they were of almost cosmic significance. Thus, neglect of suffering and overindulgence, including self-indulgence, co-exist; and people become ever less able to distinguish between tragedy and inconvenience.

In parallel with the demand for parity in treatment, there is a demand for parity of thought, feeling, and judgment (or lack of it) when we meet or consider people with psychological disturbances. They are never to be thought of as responsible for their own condition or situation, but rather as victims of something exterior to themselves (dysfunctions of their brain in this instance to be considered as exterior to themselves, that is to say to their true selves). I am reminded of what Edmund, the wicked scheming son of the Earl of Gloucester in King Lear, says:

> This is the excellent foppery of the world, that,
> when we are sick in fortune – often the surfeits
> of our own behavior – we make guilty of our
> disasters the sun, the moon, and the stars: as
> if we were villains on necessity; fools by
> heavenly compulsion; knaves, thieves, and
> treachers by spherical predominance; drunkards,

liars, and adulterers, by an enforc'd obedience of
planetary influence; and all that we are evil in,
by a divine thrusting on…

Instead of astrology, however, we believe in psychology, of whatever school – and call it progress.

This determination not to judge is compounded of three intertwined things: one a fear, the second a wish, and the third a hope.

We fear to judge because, in judging, we may overdo it, and blame those who are truly not blameworthy. More than that, we fear to appear censorious, as if there were nothing between complete and consistent moral latitudinarianism and Dickensian hypocrisy. Since the latter is now the only sin, it is best to make no judgment at all. Thus there can be nothing morally to choose between the disordered conduct of a person with a brain tumor or dementia on the one hand, and a person who has intoxicated himself with drugs on the other. If the outcome is more or less the same, the causes must be more or less the same, ontologically, and therefore morally. By not judging, we avoid the possibility of error – at least the error of blaming the victim – though not that of exculpating the perpetrator.

Our fear of appearing censorious is accompanied by the wish to appear understanding. To understand all is to forgive all; therefore, if we forgive all, we understand all. We thus put ourselves in the position of an all-merciful deity, the very deity whose existence we often are at pains angrily to deny. Suffering of any kind, even that which would once have been deemed by most people as self-inflicted, is *ipso facto* evidence of victimhood. Our philosophical motto is not I *think, therefore I am*, but *He suffers, therefore he's a victim.*

We need everyone who suffers to be a victim because only thus can we maintain our pretense to universal understanding and experience the warm glow of our own compassion, so akin to the warmth that a strong, stiff drink imparts in the cold. Of course, if it really were true, and we really believed that all suffering was evidence of victimhood, then such suffering would be a brute fact, like the height of Mount Everest or the capital of the Czech Republic; but then, as Emerson said in one of his brief excursions into comprehensibility, foolish consistency is the hobgoblin of little minds.

The purpose of such all-encompassing understanding, other than moral self-aggrandizement, is the evasion of one's own moral responsibility; for it follows that if no one is to be judged (because to judge is to judge harshly), then one is not oneself to be judged – not even by oneself. This, in effect, means carte blanche to do as you feel like, because all behavior is put on an equal moral footing: it is only to be *understood*. That is why it is not uncommon nowadays to hear someone say "I'm learning to forgive myself" (usually under the guidance of a therapist), as if such learning were hard and valuable work equivalent, say, to learning the subjunctives of the verbs of a foreign language. According to more traditional ways of thinking, learning to forgive oneself is learning how to act without scruple, how to forge ahead without regard to other people, how – in effect – to become a psychopath. Incidentally, multiculturalism has the same logical consequence. If one can claim that one's own bad behavior is part of a general cultural pattern, then that is the end of criticism; for all cultures are equal, and there is not in any case any extracultural moral Archimedean point from which to judge them. If multiculturalism means I have to accept your ways without comment, it also means you have to accept mine without comment. Like love, multiculturalism means never having to say sorry.

People have not only to learn to forgive themselves if they are to lead happy and fulfilled lives, they have to learn to *love* themselves, and puff their chests out with self-esteem. In this, at least, all therapies are agreed, of the psychodynamic, talking, counseling, and behavioral varieties. The theory is that, by believing oneself to be a bad, weak, foolish, generally incompetent, or valueless person, one sets oneself up for perpetual failure, allowing life to use one as a doormat.

A cartoon in the *New Yorker* once captured the absurdity of this kind of thinking with admirable concision. A therapist of the Adlerian persuasion (Alfred Adler, if you are old enough to remember the days of his fame, made the drive to power rather than sex the universal motive, and coined the concepts of inferiority and superiority complexes to explain the pathological variations in that drive) says to his patient on the couch, "The problem is, Mr. Jones, you really *are* inferior."

CHAPTER FIVE

Self-esteem and self-love, in the modern psychology, act as the categorical imperative in Kant's moral philosophy; or to change the analogy slightly, they are like the grin of the Cheshire Cat when all other personal characteristics have been stripped away and have disappeared. Whatever else you may do, you must always love and esteem yourself, otherwise you are doomed to a life of sterile self-denigration. In dynamic psychotherapy one must uncover the roots of a lack of self-esteem: for example, early in life when your mother did not love you enough, criticized you all the time for making a mess, derided your efforts at drawing, etc. In behavioral psychotherapy a lack of necessary self-esteem is the result of a vicious circle of thought in which reflections upon failure lead to real failure, which lead to future reflections upon failure, and so on *ad infinitum*. The object of cognitive behavioral therapy is to break the vicious circle, thus transforming a wretched mouselike creature who barely dares leave his mouse hole into a go-getter who wins friends and influences people. It is not difficult to see the connection between these ideas and the modern pedagogic tendency to praise children for their efforts, however desultory. In case people think I am exaggerating, let me here remark that an eminent professor at one of Britain's foremost institutions of higher learning, at which many Nobel Prize winners both studied and taught, informed me recently that he was not permitted to use red ink in marking his students' essays (they still wrote them by hand, apparently, to cheat by computer being too easy for them) because red ink is deemed by those in charge of the students' well-being to be too intimidating. The sight of red ink on the pages of their immortal prose might cause the little ones to lose their self-esteem, traumatize them, and mean a blighted life for ever after. Does one laugh? Does one cry? Does one despair? Does one leap for joy at such delectable absurdity? Or all or none of the above? As the Habsburg military used to say, the situation is catastrophic, but not serious.

The notion of self-love or self-esteem is in itself either ridiculous or repellent. No one ascribes his good character or successes in life to an adequate fund of self-esteem. No one says of any human achievement that it was the fruit of self-esteem. Indeed, a dose of self-doubt is, if anything, more likely than self-esteem to lead to the effort necessary (but not sufficient) for such achievement. Self-doubt, within reason, is something to be overcome; self-esteem is complacency elevated to an ontological plane.

No sensible person thinks for a moment about whether he loves or esteems himself. To do so is inherently a form of vanity, but much worse than mere physical vanity (vanity about one's clothes, etc.), which at least has an other-regarding, which is to say social, quality. Self-esteem is vanity not about appearance but about the very fiber of one's being, about one's character. To have it is to award oneself a medal, as it were, merely for existing, for not taking the only alternative to existence, namely suicide. For most people, the failure to commit suicide is an easy hurdle to jump.

But even if we were to grant self-esteem some value as a concept, either explanatory, descriptive, or as a universal desideratum for human beings, it would still be wrong to disconnect it entirely from moral considerations. Anyone who thinks that self-esteem is a good in itself and not potentially a manifestation of the deadliest of the deadly sins, pride (deadliest because all the others may be derived from it), should be put to reading *Coriolanus*, wherein the consequences of an excess of self-esteem are laid out to the moral enlightenment of theatergoers. Coriolanus thinks excessively well of himself because of his noble birth, that is to say, almost *ex officio*. While such self-esteem makes him brave, it also causes him to forget that bravery is one of those virtues that cannot be free-standing, and that to be truly a virtue, it must be exercised in pursuit of a worthy goal; his excessive self-esteem also causes him to denigrate others, likewise *ex officio*, for not being noble, without examination of their individual qualities. He eventually allows his humanity to overcome his self-esteem, and he dies for it; but he would never have done so but for his infernal self-esteem.

If it were possible to have too little self-esteem, it seems almost a logical consequence that it is also possible to have too much of it; but which any person has – too much or too little (or just enough) – irreducibly requires a moral judgment, an estimate of what is appropriate, and this is a judgment that cannot be internal to psychology alone, but must entail a notion of the good, both for the individual and for society. If, for example, the increase in a particular person's self-esteem were severely damaging to others, one would not advocate it, even if it were beneficial for the person himself. It is no good saying that self-esteem is not *real* self-esteem if it harms others, because no one can derive self-esteem from harming others; that is to smuggle a moral judgment into the notion, which as a purely scientific concept was supposed to be free from it. As a matter of mere observation, moreover, plenty of very bad people are full of self-esteem: notorious criminals, for example, who frequently take great pride in their exploits. The self-esteem of the evil is particularly chilling.

It is therefore entirely possible that those who complain of low self-esteem have a nearly correct estimate of themselves (it should, of course, be a little lower for having thought of self-esteem in the first place). The concept of self-esteem is one that reeks of bad faith, of special pleading on behalf of oneself. When as a doctor I replied to someone who complained to me of his low self-esteem that at least he had got one thing right, he would laugh instead of becoming angry, as he would have done if he had truly and sincerely believed that his problem was that of low self-esteem. In his laugh was an implicit acknowledgement that he knew all along he was trying to deceive himself as well as others. A man may despise himself for being as he is, but that does not absolve him of the responsibility for being as he is. There may indeed be extenuating circumstances, and on occasion complete excuses, though it is for others to assess them; Man cannot extenuate away his responsibility for himself as he so often tries to do with the encouragement of psychology.

To reiterate Edmund in *Lear*: "An admirable evasion of whoremaster man, to lay his goatish disposition to the charge of a star!"

Self-esteem is a concept that belongs to the psychology of the Real Me. The Real Me, of course, is someone who is inherently good and admirable: Man being by nature good, inside every bad man there's a good one trying to get out, obstructed, alas, by such phenomena as low self-esteem. The Real Me may actually have no obvious connection to the Me as it acts in the world and appears to others. It is a secret and beautiful garden often accessible only by means of psychology.

Now it is perfectly true that all of us sometimes act out of character: a perfectly consistent character has probably never existed, and we might find him disconcerting if we met such a one. Thus a sweet-tempered person may lose his temper on occasion without losing his reputation for sweet-temperedness. But a man who regularly loses his temper on the slightest provocation cannot claim *really* to be sweet-tempered because he knows that in the secret garden of his Real Me he is patience personified. The doctrine of the Real Me, who has nothing to do with and no resemblance to the Phenomenal Me, that is to say the Me who eats, drinks, and sleeps, is yet another admirable evasion of whoremaster man, for it allows us to do as we please without having to think badly of ourselves, to experience genuine remorse, or even to examine ourselves honestly. This is because the verdict is always decided in advance: we are always, in the innermost recesses of our being which the Real Me inhabits, innocent. And, of course, the innermost recesses are much more real than the outer being, which is as superficial as lipstick. In my childhood I used to cross my fingers when I told a lie, which allowed me to believe that the lie was not really a lie because I knew, deeper down than what came out of my mouth, the truth. In effect, we are always in a state of inner emigration (as those opponents of Nazism who stayed in Nazi Germany called it) from our phenomenal selves.

There is nothing new in this, no doubt; there rarely is anything new where humanity, or for that matter where inhumanity, is concerned. The question is rarely whether something is new under the sun, but whether it is more, or less, prevalent than it was. And the psychology of the Real Me certainly would, in logic, encourage the evasions of whoremaster man.

Needless to say, the Real Me is wheeled out for use only in the context of bad or illegal behavior. No one claims of his good deeds that they are an exception from his bad character, that (for example) he would usually snatch an old lady's purse rather than help her across the road. It is only bad behavior that is exceptional and in need of psychological explanation. All good behavior is perfectly consonant with the Real Me, and is therefore not at all mysterious. On the contrary, such behavior is the expression of the natural self, which is blocked by some pathological process or other. The hold that the psychology of the Real Me has on humanity explains why so much more effort has gone into explaining evil conduct than good. The alien is always more interesting than the familiar, the abnormal than the normal.

The Real Me, then, shines diamond-bright, or would do so if only the dross of the Apparent Me were cleared away, the way diamonds are found in the workings of

South Africa. The psychologist or psychiatrist is the miner, the dissatisfied or bad person (that is to say, the person who does bad things that he at least pretends he would rather not do) the mined.

A natural corollary of the Real Me is the Real Him, the true person beneath the veneer of his actual behavior. Unlike the Real Me, who is always good, the Real Him can be good or bad, depending on whether or not the person who is seeking the Real Him likes the Apparent Him. It is more usual, however, to attribute good character to those who behave badly than bad character to those who behave well (the latter propensity often being the consequence of envy). I once heard a fond mother of a boy aged fifteen, who had burgled more than two hundred houses, say of him on the radio that "he's a good boy really," that is to say, a lad with a heart of gold, despite the considerable amount of misery to others that each of his crimes had almost certainly caused. No doubt it is a natural and to some degree necessary thing for a mother to indulge in special pleading on behalf of her son, but it is absurd that it should be accorded any intellectual respect.

The search for the Real Him is now a reflex among intellectuals when faced with the conduct of others that they find reprehensible; for they are uneasy with the categories of good and bad, of whose metaphysical foundations they are so uncertain, and by which the great mass of mankind (from which intellectuals must distinguish themselves) judges men and events. For example (and a typical example), an article in *Le Monde* in June 2014 began "Who knows the real Mehdi Nemmouche?" The reason anyone would want to know the *real* Mehdi

Nemmouche is because the *apparent* Mehdi Nemmouche had just been caught with weapons in his possession with which he had shortly before shot four people dead at the Jewish Museum in Brussels, along with the camera with which he had hoped to record the killings for posterity. He was trying to escape to Algeria when he was apprehended.

The article's headline was "The Multiple Lives of Mehdi Nemmouche." It started:

Who knows the real Mehdi Nemmouche? Suspected of being the person responsible for the killing of four people at the Jewish Museum in Brussels on 24 May, he has several lives, without anyone knowing how exactly they are connected.

To help us understand and uncover the *real* Mehdi Nemmouche, now 29 years old, one of his childhood friends was quoted as saying that he was "a quiet boy, discreet, not at all aggressive, and who never had any problems with anybody." Another school friend said, "He was a good pupil, he never had any fights, had friends, dressed normally in jeans and sneakers..." But for the criminal justice system he was a young man with a "solid history of delinquency," with no fewer than seven convictions in ten years, many of them for robbery with violence. He spent five years between 2007 and 2012 in prison, in which he was "radicalized," that is to say, he was given (and adopted) an ideological justification for his psychopathic behavior.

So who is the *real* Mehdi Nemmouche, the quiet, well-behaved boy or the delinquent young man who shot four people with callous and prideful indifference?

The question could only be asked by someone who thought that the answer mattered, and that furthermore there was a reality that lay deeper than the superficial act of shooting. If, for example, we discovered that Nemmouche had always been nice to his aged grandmother, should we conclude that the *Real Him* was a person with warm and generous feelings for old ladies?

Everyone has more than one facet to his character and conduct, all of them equally real. Hitler was nice to his dog and liked receiving flowers from small children. One cruel act is just as real as a hundred kind ones, and vice versa; which is the more significant is an irreducibly *moral* judgment, which no amount of psychological enquiry can help us to make. In my clinical work, I discovered that many violently jealous men strangled their consorts, and not for the sexual pleasure that asphyxia is sometimes said to confer on the asphyxiated.

"Does he ever strangle you?" I would ask my patients who had violently jealous male consorts.

"Yes," more than one of them replied, "but not all the time, doctor." (One had even asked her consort not to strangle her in front of the children.) I could not always get them to see that a minute of strangulation was more significant than an hour without: but at least it was their decision to take. But if they were killed, it would be no defense of the killer that they had accepted his previous strangulations, or that he had spent the vast majority of his life not strangling anyone.

The doctrine of the Real Him is a watered-down secular version of Christian redemption, with Man in the place of God. Inside every person there is a core of goodness that is more real, more fundamental, than any evil act he might have committed, and which it is the purpose of punishment to bring to the surface. That is why punishment is now believed, by all right-thinking people at least, to be therapeutic, which is to say redemptive, in purpose and intention. The European Court of Human Rights recently ruled that whole-life sentences to prison are against Man's fundamental rights because they eliminate the possibility of repentance and redemption (known in the trade as rehabilitation). Thus, the judges of a court that is supreme in matters relating to supposed human rights for a continent on which, within living memory, tens of millions of people have been systematically starved or abused to death or put to death industrially on an unimaginably vast scale, could conceive of no crime so terrible that the person who committed it was beyond earthly redemption. On this basis someone like Himmler, had he not committed suicide, or Beria, had he not been shot by his erstwhile colleagues, would have been eligible for parole, provided only that they showed themselves reformed characters by, for example, making toys for children or Braille books for the blind. (A serial killer really did once upbraid me in print for suggesting that he – who three decades earlier had kidnapped at least five children, sexually abused and tortured them to death, then buried them in a remote place in the moors – should never be released from prison, on the grounds that he now spent much of his time making Braille books for the blind, which was more, he claimed, than I had ever done to help anyone. In other words, he had redeemed himself, and canceled out the torture and murder of five children, by subsequent good works, thus expressing the *Real Him*; and, in the words of the commonly used cant phrase, he thereby *had paid his debt to society*, as if good and evil were entries in a system of double-entry book-keeping, so that if one did enough good works in advance, one would have earned the right to torture and murder five children.)

Men can change; this is their glory and their burden, for it is precisely the capacity to change that renders them responsible for their actions; but what they do may be irreparable.

CHAPTER SIX

The notion of rehabilitation, to which the European Court of Human Rights apparently attaches so much importance, implies that those who do wrong do it because they are ill and in need of moral physiotherapy, equivalent to the exercises that people who have broken their hip or had their hip replaced must do in order to be able to walk again. The idea that no man does wrong knowingly (for if he knew what was right he would automatically do it) has been replaced by the idea that no man does wrong healthily, that wrongdoing is a kind of disease to which there is a cure. C. S. Lewis criticized the idea of punishment as therapy as emptying human life of all its specific significance in his incisive essay "The Humanitarian Theory of Punishment"; moreover, therapeutic punishment is compatible with both the most absurd leniency and the most revolting severity. As if this were not enough, it is also incompatible with the rule of law, for therapy must continue until it is successful; and when and whether it has been successful must always remain speculative. The length, duration, and severity of therapeutic punishment will depend upon prognosis, a difficult and uncertain art at the best of times; a man is to be punished for what he *has* done, not what he *might* do in the future. To employ prognosis in the determination of the length or nature of individual sentences is to make the law arbitrary in its workings and dishonest in its pretensions. Thus psychology, in practice, undermines the rule of law.

The intrusion of psychology into law on the therapeutic theory of punishment, much beyond the establishment of clear-cut madness that has for centuries been a complete excuse or strongly extenuating circumstance, goes ever further. In France as I write this there is a projected law requiring judges to speculate on the most effective punishments for the criminals who come before them. It is inevitable that they will sometimes be dupes, sometimes relatively (and therefore unjustly) harsh; for they will forget Kent's warning to Lear that Cordelia's absence of effusiveness is not necessarily a sign of absence of love:

> *Nor are those empty-hearted whose low sound*
> *Reverbs no hollowness.*

The new law is one of a series known as the Law on Recidivism, a title that makes the therapeutic aims of the criminal law perfectly plain. The object of the law is to *heal* the criminal so that he sins no more; but only a moment's reflection is necessary to realize that, where there is a choice of incarceration and so-called community sentencing, a reduction in the rate of recidivism is perfectly compatible with a rise, even a huge rise, in the numbers of crimes committed; and vice versa, with a rise in the rate of recidivism coincident with a fall, even a dramatic fall, in the rate of crime. The criminal law is supposed to be protective of the public, not curative of the criminal (even supposing that there is something wrong with him in the first place that requires "cure," a proposition that undermines his humanity, as C. S. Lewis pointed out). If in protecting the public the rate of recidivism were to decline, that would be all to the good, of course; but it is the protection, not the recidivism, that should be aimed at. The fact that so elementary a deformation of the law can be entertained very widely, especially by influential intellectuals, only goes to show how dominant psychological modes of thought have become, to the great detriment of clarity and true understanding.

The psychology of the Real Me also leads naturally enough to that strange form of *langue de bois* known as psychobabble. In this debased and filleted language, people talk endlessly of themselves without true revelation of anything about themselves. It is a manifestation of self-absorption without self-examination, and it aims to create the impression of frankness or openness without the need of honesty, which is always painful. One person speaking psychobabble sounds much like another, which is not surprising, because one of the effects (if not the object) of such language is precisely to evade talking of those experiences and thoughts that constitute human individuality. To reveal your thoughts is to make yourself vulnerable, for by doing so you expose to others what is unique to yourself, the very fiber of your being as it were; and to do so may lead others to reprehend, disdain, ridicule, or despise you. That is why true candor is so difficult to manage; there are times, places, and interlocutors for it. Just as Voltaire remarked that the best way to be a bore is to say everything, so it is the best way to be a fool (and often a boor as well). Confidences are not confidences that are told to everyone, but the problem with true confidences is that they require trust in the confidante: and whether a confidante is worthy of the trust placed in him is a matter of discrimination and judgment. People abjure judgment for fear of error and appearing ridiculous in the eyes of both themselves and others; better, then, not to exercise it, and to adopt instead an undiscriminating but unrevealing openness, a simulacrum of candor. Perhaps this helps to explain the phenomenal success of Facebook and the like: the new social contract being that I pretend to be interested in your trivia if you pretend to be interested in mine. By this means one expresses oneself, or at least communicates something without the painful necessity to think or reflect on anything. But as habits become character, so the habit of superficiality eventually becomes (paradoxically) deep, or at least deeply ingrained. Perhaps this explains the increasing need of extravagant expressions and gestures that seem to accompany thinness of content. Only thus can one obtain notice in the torrent, the ocean of verbiage, though such extravagance of expression and gesture is ultimately futile, since it leads to a competition for attention that no one can win.

A more recent form of psychobabble is that which relates to the arcane science of neurochemistry, which has turned into another powerful tool in Man's eternal search to evade his responsibilities. I do not mean here in any way to decry neurophysiology and its accomplishments. The determination and human genius that have elucidated the anatomy, physics, physiology, and chemistry of the transmission of nerve impulses are astonishing.

When I consider how it was discovered that nerve cells are separate from one another and communicate by means of discharges of chemicals, I am lost in admiration. But what begins as a scientific advance quickly ends as an urban myth. A technical advance, an increase in knowledge, becomes the explanation of and key to human existence as a whole, even if the scientists responsible for the advance make no such large claim (though sometimes they do). The notion of an imbalance of brain chemicals as the origin of one's thoughts, desires, mood, and behavior, especially when bad, was accepted by educated people of the late twentieth century with a credulity not exceeded by that of medieval peasants in the face of religious relics, though with less beneficial aesthetic – and possibly psychological – results. And what starts as a fashion among the educated soon filters its way down to the uneducated.

It is now not uncommon to overhear people talking of their own brain chemistry as if it were the dangerous reaction of potassium permanganate and hydrochloric acid in a test tube (the danger being the generation of chlorine gas, much as brain chemistry gives off bad temper or selfishness). They speak with the strange authority of the religiously certain, the imbalance of their brain chemistry being as evident a fact to them as any physical object in their immediate surroundings, or as the existence of God to the religious. I even overheard one young woman on a bus telling her fellow passenger that her problem was that she lacked the right amount of lithium in her body for her brain to function normally, and it had to be supplied in the form of pills, much as those lacking thyroid hormone need to take thyroxine. Lithium carbonate is indeed given to patients suffering from manic depression, to try to even out their excessive moods; but this is not because they are for some reason – nutritional, metabolic, or other – deficient in lithium, any more than a person suffering from pneumonia is deficient in penicillin.

The popularity of brain chemistry as the explanation of all human behavior – at least all human behavior that, because of the difficulties it causes, seems to stand in need of explanation – began in the 1980s with the extremely successful marketing of new drugs, supposedly antidepressant, called selective serotonin reuptake inhibitors (SSRIs). So successful has their marketing been, indeed, that at any time about a tenth of the adult population is taking them.

Now these drugs were sold on a theory that is by no means new, namely that mental depression is caused by an insufficiency of one or another of the known neurotransmitters, the chemicals released from the end of a nerve cell to activate (or, in some cases, to prevent the activation) of an adjoining nerve cell. The known neurotransmitters are not very many in number, unlike nerve cells themselves, which are numbered in the billions, each of them with so many connections that the brain can hardly conceive of its own complexity.

The psychiatrist Joseph Jacob Schildkraut first popularized (at least among the medical profession) the notion of a biochemical cause of depression in 1965, in a paper titled "The Catecholamine Hypothesis of Affective Disorders." He was not the first to put forward the hypothesis, but he synthesized and drew together the evidence, such as it was, very succinctly. In those days the favored candidate for the cause of depression was a deficiency in noradrenaline rather than serotonin, for the first effective antidepressants, such as imipramine and amitriptyline (whose mood-elevating effects were first noticed quite by chance), as well as the monoamine oxidase inhibitors, seemed principally to have an effect on the former neurotransmitter rather than on the latter.

There were many difficulties with the hypothesis from the very first. For example, the antidepressants worked, when they did so, only after two or three weeks, while the pharmacological effect on monoamine metabolism was immediate. In any case, the biochemical theory of mood regulation hardly spilled over into the general population, and this for two reasons: first, it remained an arcane medical matter discussed in learned and inaccessible journals; and second, depression of a severity thought to require treatment was then a relatively uncommon condition, more or less coterminous with melancholia. This had all the marks of a genuine illness. It often struck for no apparent reason; there was a strong hereditary predisposition to it; it could be reproduced in its symptomatology by certain undoubtedly physical illnesses; it had physical accompaniments and even, just possibly, a physical marker; there were strong, uncharacteristic, and unjustified feelings of guilt unassuageable by anything except successful treatment or natural remission; it was, or could be, very severe, and if untreated could end in complete stupor or the bizarre nihilistic delusions of Cotard's syndrome, in which the patient believes he is rotting away, that his body is gangrenous, that he is dead already; and so forth. The historical record of lunatic asylums show that, before there was any effective treatment of such melancholia, patients would be kept under close observation to prevent them from killing or starving themselves until they recovered spontaneously – which they usually did, but only after a prolonged time of deep misery. In one medical memoir published in 1921 by an asylum doctor, Montagu Lomax, I read that melancholic patients during the day would be lined up against a wall on chairs, a table pushed up to them to prevent them from going anywhere, and an attendant would sit opposite them to observe them: a form of "treatment" that continued until their condition abated naturally.

Patients such as these were few and unmistakable. The new antidepressants were undoubtedly effective, which was a great relief to all, for the only other effective treatment at the time was electroconvulsive therapy, now carried out in a more refined way, but in those days a frightening and degrading procedure; but they, the antidepressants, had a number of troublesome side effects, not least among them a propensity to send patients to the opposite end of the mood spectrum, that is to say, they became manic.

Because they were so few, the cure or amelioration of such patients had little cultural impact. Nevertheless, the ground had been prepared. In its article on Dr. Schildkraut, who died in 2006, *Encyclopædia Britannica* says:

He was widely known for his research paper "The Catecholamine Hypothesis of Affective Disorders," published in the American Journal of Psychiatry in 1965, which helped establish a biochemical basis for depression and other mood disorders.

In other words a hypothesis, though unproved and now discarded by almost everyone (at least as to the bio-chemical substance supposedly disordered), helped "to establish a biochemical basis for depression."

Two developments were necessary for the "chemical imbalance" theory of the difficulties of existence to become so widespread as to seem self-evident to half the population (the better-educated half at that): first, the loosening of the diagnosis of depression to encompass almost all forms of human unhappiness; and second, the development of new antidepressants, the famous – or infamous – SSRIs.

The first of these conditions has been accomplished so thoroughly that the very word *unhappy* has been all but eliminated from common parlance. For every person whom you overhear talking in a public place of their unhappiness, you hear at least ten talking of their depression. Few are those who now admit to being unhappy rather than depressed, another "admirable evasion," for since depression is an illness – caused, of course, by a chemical imbalance – it is only natural that those who suffer from it should seek medical treatment whenever they experience any deviation from happiness, which is the natural state of Mankind, as well as its inalienable right (the finding having replaced the search as the inalienable right). If someone admits to unhappiness, it might be that his own ill-conduct, foolish or immoral, has contributed to it; but if he is depressed he is the victim of an illness, of something which, metaphysically speaking, has fallen from the sky. This makes of the person an object rather than a subject; and while there *are* circumstances in which a person is a pure and unadulterated victim, an object, as it were, they are relatively few where the workings and content of his mind is concerned.

The patients have been assisted in their evasions by the psychiatric profession, with members of which they now have a dialectical relationship. The psychiatrist diagnoses depression without any reference to the circumstances or recent history of the patient. As one eminent psychiatrist of my acquaintance put it, there is only depression and more depression – that is to say, no difference between mild dysphoria and melancholic stupor. They, the psychiatrists, have a checklist of symptoms; and if patients claim to suffer from a sufficient number of them (not very many, in all conscience), they get the pills. Some of the symptoms have an irreducibly moral content, self-esteem for example, the loss of which is invariably pathological in the eyes of the psychiatrists; or a feeling of guilt, an increase in which is likewise pathological, irrespective of any possible justification for it. The way in which the *Diagnostic and Statistical Manual of Mental Disorders* of the American Psychiatric Association deals with bereavement (its one exception to the rule that the actual circumstances of the possibly depressed patient are not be taken into account in diagnosis, only his symptoms) would be comical were it not the manifestation of a deeply impoverished notion of human life, as well as very important in its practical, social, and cultural effects. It allows that the symptoms of depression are not to be counted as depressive illness in the first two weeks after a bereavement, but thereafter become a depressive illness. On this view, Tennyson's great poem, *In Memoriam*, is but the morbid product of depression (except for any stanzas written in the first two weeks after

Arthur Henry Hallam's death, which must have been few, since the poem as a whole took him seventeen years to write). How much better it would have been if Tennyson had been given an SSRI to get him over the loss of his friend! English literature might have been one long poem the less, but Tennyson would have been much the happier. On this view, also, Hamlet's great outburst in protest at his mother's swift remarriage to Claudius would be not so much meaningless as psychiatrically primitive:

> A little month, or ere those shoes were old
> With which she followed my poor father's body,
> Like Niobe, all tears. Why she, even she –
> O God, a beast that wants discourse of reason
> Would have mourned longer!

Not so! A creature now possessed of discourse of reason and mourning longer than two weeks would have gone down to his doctor and been prescribed a little Prozac. Gertrude wouldn't then have married Claudius to get herself over her grief, Hamlet would have ascended to the throne, Ophelia would have been queen, and Polonius, Ophelia, Gertrude, Claudius, Laertes, and Hamlet would have survived.

If only Hamlet had listened to Prozac – or, more accurately, had had Prozac to listen to! Listening to Prozac, you may remember, was a best-selling book published in 1993 by the psychiatrist Peter D. Kramer. Prozac was the first of the SSRIs to be marketed, and its claimed advantages were many. It was supposedly effective, with fewer side effects and less dangerous in overdose than the earlier antidepressant drugs known as tricyclics. The latter caused a number of disagreeable and potentially dangerous side effects, such as drowsiness, dry mouth, constipation, excessive sweating, low blood pressure on standing (particularly important in old people, for whom a fall is often the beginning of their decline to death), and irregularities of the heartbeat, sometimes fatal after an overdose (which of course depressed and suicidal patients are more inclined on average to take).

If the claims made on behalf of the SSRIs had been that they were as effective as the older drugs in the treatment of melancholia, but were safer and less disagreeable to take, they would have represented an advance, not epoch-making like the discovery of anesthetics or antibiotics, perhaps, but worthwhile nonetheless. But Dr. Kramer's book suggested much more than this, that in fact Prozac could amend a defective personality, or one that the person who had it perceived as defective. So great was our knowledge of and command over neurotransmitters, implied Dr. Kramer, that we were entering the age of cosmetic neuropharmacology, such that we would be able to design our own personalities: a little more self-confidence here, a little less irascibility there. We could be exactly who we want to be, not by the traditional means of discipline and self-control, but rather by taking a judicious mixture of pills. This, of course, is just what whoremaster man wants to hear. The only thing that has changed since he made guilty of his disasters the sun, the moon, and the stars is that he now makes guilty of them noradrenaline, serotonin, and gamma-aminobutyric acid.

This neurotransmitter reductionism, in which an over- or undersupply of a handful of biological substances in the brain supposedly accounted for all our disasters, should have been laughed out of court on commonsense grounds. To iron out thus and simplify the enormous complexity of our lives, to reduce it to the interplay of the few neurotransmitters known, was to make nineteenth-century phrenology look sophisticated by comparison. Einstein, remember, said that explanations should be as simple as possible, but not simpler than possible; and perhaps the desire for the illusion of understanding is generally greater than the desire for understanding itself. And people believed because they wanted to believe: at last the hard business of living was over; from now on it would be happiness and success all the way.

At any rate, rumors of great advances in understanding soon seeped out into the general population, which then became avid for self-improving drugs.

As it happened, practically all the claims made on behalf of the SSRIs were not only false, but deliberate lies, though some of them lies only by implication (the most effective kind, not only because they can be disavowed when exposed, but because the implicit is always more effective than the explicit). It was true that they were safer than the old drugs, though by no means completely safe, and generally had fewer side effects than the old; but it turned out that they had side effects not initially apparent – for example, on withdrawal of the drugs – knowledge of some of which was suppressed by the manufacturers.

Nor were they nearly as effective as antidepressants as they were once claimed to be. Among the means used by the pharmaceutical companies to conceal their ineffectiveness, selective publication of the results of trials was a great favorite. The companies would sponsor trials of their product, but if the results were negative or unimpressive they would not publish them, publishing only the results of those trials that were favorable. By this means, in theory, almost anything could be proved to cure any condition whatever, provided only that enough trials were performed: for one or some of them would yield positive results purely by chance; and if the results of the trials with negative results were suppressed (that is to say, deliberately not published), while the results of the positive trial was bruited abroad, a misleading impression of the treatment's efficacy would be created. And so it was with the new antidepressants, if not in quite so extreme a manner. Doctors were systematically misled by pharmaceutical companies into believing that SSRIs were more effective than they are by means of the suppression of unflattering data.

Never mind that they scarcely worked: they sold phenomenally. A tenth of the adults in many Western countries now takes these drugs, and recently I saw that in some towns in England, where the unemployment rate remains stubbornly high years after deindustrialization, it is as much as one-sixth. Here one can sympathize not only with the people themselves, but with the doctor who must look after them. Such people are often not highly educated, not of exceptional intelligence, and their social world has been smashed up both by circumstances and their own bad choices; they may have few skills or economic prospects, no intellectual, political, or religious interests or beliefs, and only the crises brought about by social pathology can distract them from, or give meaning and variety to, the bleakness of their existence. Almost as despairing himself, the doctor gives them pills for lack of anything else to do; and fortunately for him, the rumor has reached even those benighted parts that unhappiness is a pathological state brought about by chemical imbalance in the brain. One might define Man as the only animal susceptible to the placebo response, provided it is remembered that the placebo response of the doctor is often greater than that of the patient. With how much relief does he feel that he has done *something* in prescribing a drug that he knows perfectly well has but a faint chance of success!

The number of conditions for which the SSRIs were prescribed, some of them specially invented to fit the drug, gradually increased until it almost seemed that the whole of human misery was accounted for by errors of cerebral serotonin metabolism. Unhappy? Given to eating in binges? Shy at parties? Take Prozac, or one of its analogues.

It is hardly surprising, then, that with every person in the Western world (according to the *Diagnostic and Statistical Manual*) suffering on average two psychiatric disorders a year, and with the uncritical acceptance of the biochemical imbalance theory of mental disturbance, that a fifth of the population of the United States is now taking psychotropic medication of one kind or another. This might be described as the triumph of marketing over the awareness of the tragic dimension of life: that dissatisfaction is the permanent and ineradicable condition of Mankind because of the incompatible and conflicting desires in every human breast. As for children, it is never too early to drug them up.

CHAPTER SEVEN

There is, I trust, no need to point out again the convenience of the neurochemical hypothesis of unhappiness and undesirable conduct to those who seek to "lay their goatish disposition to the charge of a star." In other words, it's not me, it's my neurotransmitters.

Or possibly my genes. From time to time, with a great fanfare, scientists make announcements that the gene for this or for that behavioral trait has been found: the gene for aggressive behavior, for miserliness, for sexual promiscuousness, for alcoholism or drug addiction, and so forth. Then the experimental work fails to be reproduced, though this failure is given much less prominence in the press than the original fanfare, leaving the public with the impression that great progress in the explanation of human behavior has been made. This impression is strengthened by the evident fact that people are born with a certain temperament, for example exuberant or reserved, and that this is sometimes clearly inherited, whether genetically or culturally, from their parents.

It would be surprising if genes had no influence on character, and therefore on behavior. Man is programmed to learn language, for example. But being programmed to learn language is not the same as being programmed as to what to say, upon which many influences have been brought to bear, including those unique to the individual who speaks. The fact that language is rule-bound does not mean that the number of things that can be said is finite, nor does the fact that some statements are incomprehensible or meaningless reduce the possible number of meanings that can be expressed in language. "Whereof one cannot speak, thereof one must be silent," said Wittgenstein; but the infinitude of what can nevertheless be said, and actually is said, is one of the reasons for believing that no full explanation of human behavior will ever be found, at least if it is granted that what men say is an important part and even a determinant of their behavior.

Let us take drug addiction as an example. There seems to be a slight genetic propensity toward it, as revealed by studies of identical twins as compared with other siblings. But this cannot possibly explain the huge variations over time in the rate of drug addiction. In Britain, for example, the rate of addiction to heroin was once so low that, as late as 1966, it seemed not a serious problem at all to Lord Brain, the great neurologist who was charged by the government with investigating its extent and making recommendations as to policy. A system existed for the registration of such addicts who, once registered, were eligible for the free prescription of heroin, paid for from general taxation. In the 1950s, fewer than a hundred such registered addicts were known; and while, of course, this might have been an underestimate of the true numbers, it was unlikely to be a gross underestimate, since addicts had every motive to reveal themselves. In Britain today there are between 250,000 and 300,000 heroin addicts, half of whom inject and the other half of whom smoke the drug. Genetics will not throw much light on this change, even if it explains in part (and only in part) why person A rather than person B, in similar circumstances, listens to the siren song of the drug.

Likewise, in the United States genetics are powerless to explain the sudden rise in the number of deaths from overdoses of prescription opioids, usually in conjunction with other drugs – fivefold from the turn of the millennium; why in New York City the rate of such deaths rose from 0.39 per 100,000 of the population in 1990 to 2.7 per 100,000 in 2006, a seven-fold increase; or why in the same city deaths from heroin overdose rose 71 percent between 2010 and 2012.

Of course, falls have also to be explained, as do rises. After the establishment of the People's Republic of China in 1949, addiction to opium declined rapidly for the very good reason that Mao Tse-tung threatened addicts with condign punishment if they did not choose to "recover." Most of the addicts *did* choose to recover, and no research into genomes was necessary either to achieve or to explain this. In France there has of late years been a startling decline in the number of fatal road accidents, in part at least because the police have enforced the law against driving while intoxicated. Other contributory factors are better car design, better and faster treatment of trauma, and enforcement of speed limits, none of them intimately linked to the genome. In Britain, the decline – for once, a decline that is welcome – has been even more dramatic, the death rate on the roads having fallen by a half between 2000 and 2012, having already halved between 1960 and 2000, the latter despite a vast increase in vehicular traffic. Anyone who tried to explain this by reference to genetics or changes in neurochemistry would rightly be suspected of insanity. Moreover, though there was a fall of only 9 percent in per capita consumption of alcohol between 2007 and 2011, prosecutions and convictions for drunken driving in that time fell by a third, which probably reflected the actual prevalence of drunken driving in the population, since the fatal-accident rate fell *pari passu* with that of prosecutions and convictions. This in itself is instructive, for it suggests that, notwithstanding the effect of alcohol on neurotransmitters (which presumably remained the same between 2007 and 2011) and the near constancy of the population's genome, conduct under the influence of alcohol changed significantly for some other reason (perhaps an increased fear of losing a driving license in conditions of economic uncertainty), at any rate for a reason infused with *meaning*. The disjunction between alcohol consumption and conduct should surprise no one, for the fact that whole populations behave differently under the influence of alcohol, for example in their levels of drunken aggressiveness, was pointed out more than forty years ago in a lamentably disregarded book by Craig MacAndrew and Robert B. Edgerton, *Drunken Comportment*. No difference in neurotransmitters or the genome is likely to explain this; to try to analyze human behavior by reference to genes and chemicals is to regard humans in the same light as *Drosophila*, the fruit fly so favored by geneticists because it is easy to maintain and reproduces so quickly.

Periodically, with the monotony of the ticking of a clock, genetic explanations of criminal behavior come back into fashion, perhaps because criminality is so exasperatingly intractable, and it is better to have a false explanation of it than no explanation at all. I will not here deal with Émile Durkheim's theory that a society needs a certain number of criminals who serve to increase social solidarity among the rest of the population, a "them" opposed to whom we can be an "us"; suffice it to say that society always seems to have more criminals than strictly it needs, and no one, I think, has ever thanked a criminal for increasing social solidarity by means of his depredations, nor has any criminal entered such a plea in court as an extenuation of his conduct. Nor will I deal with the thorny issue of what criminal behavior actually is, a thorny issue because it keeps changing, especially nowadays when legislatures add crimes to the statute books as boys add stamps to albums, such that the average citizen can now hardly draw breath without the unwitting commission of a criminal act. The fact is that, for most people, crime still means theft, robbery, assault, and murder.

Are these activities hereditary, written in the genes? Attempts to prove them so have been legion; and an initial plausibility is given to the idea that criminality of the kind I have just mentioned is a biological fatality by the fact that it is quite clearly, and always, concentrated in a certain sector of society. Theft runs in neighborhoods, if not in families.

Cesare Lombroso, the Italian doctor and criminologist, believed that crime was hereditary degeneration and degeneracy; the whole eugenics movement, supported by such luminaries as Bernard Shaw, H. G. Wells, and the socialist biostatistician Karl Pearson, believed implicitly in the inheritability of criminal behavior. In 1929 Johannes Lange published his *Crime as Destiny: A Study of Criminal Twins* (described by J. B. S. Haldane, the Marxist geneticist, in the foreword to the English translation as "a masterpiece of scientific psychology"), in which Lange found that 13 of 17 pairs of identical twins were concordant for a criminal history, whereas only 2 of 13 pairs of nonidentical twins were concordant. In 1964, the well-known psychologist H. J. Eysenck published his *Crime and Personality*, in which he similarly alleged the predominant influence of genes on crime; and more recently it has been alleged that the comparatively high homicide rate in the United States (compared with Western European countries) is explicable by the presence in the population of approximately 10 percent of blacks, who are genetically inclined to be more violent.

No one alleges, of course, that criminality is inherited in the same way that Mendel's peas inherited their color, by means of a single gene; but in order for such complex behavior as criminality to be inherited at all – if it is – some multiplicity of genes must be involved, as yet undetermined. It is not necessary to know precisely which genes are responsible to know that a trait is inherited.

Let us examine the claim a little further. Just to remind ourselves, it is this: that the difference between the homicide rates in Europe and the United States are attributable to a much higher proportion of blacks in the United States, a race of people who are genetically inclined to commit homicide. The official statistics of homicide, incidentally, may be untrustworthy in either direction, but since everyone uses them as if they were trustworthy, I will also do so. The argument both for and against the heritability of crime is based upon such statistics.

The first thing to note is that the proportion of blacks in the population of the United States has remained more or less constant over the last century. But in 1900, the homicide rate in the United States was 1 per 100,000 of the population, more or less what it is in Britain or France today (though higher than Britain's or France's then). By 1933, however, the rate had risen to 10 per 100,000, the proportion of blacks in the population remaining constant. By 1940 it had halved again; but in the 1970s and '80s it doubled back to 10. It has since halved. Clearly these very considerable fluctuations have nothing to do with the proportion of blacks (or their propensity for violence) in the American population.

The recent reduction is, of course, welcome; but we should notice one thing more. A paper was published not long ago that estimated that, if the same resuscitation and surgical techniques were used today as were used in 1960, that is to say, well within living memory, the homicide rate would be five times higher than it is: which is to say that the homicidal attack rate is at least *twenty-five* times higher now than it was in 1900. In fact, it must be higher than that, for there was much progress in the treatment of trauma between 1900 and 1960 (in 1900, for example, there was no blood transfusion or other intravenous fluid replacement to treat cases of shock). Let us suppose what is likely (indeed, is a modest underestimate), that if the same resuscitation and surgical techniques were used in 1960 as were used in 1900, the homicide rate in 1960 would have been between two and four times higher than it was. This would mean that the homicidal attack rate today would be at least *fifty* to a *hundred* times what it was in 1900. If this is so, then any genetic influence must be swamped to the point of irrelevance by other factors, whatever they might be.

Now here let me point out that we cannot conclude from this that people in America today are a hundred times more murderous in intention than they were in 1900. This is because the very fact that people today are much more likely to be saved by medical intervention than they were in 1900 may have, indeed probably *has*, seeped into common consciousness (or subconsciousness), and thereby has altered behavior in the direction of the free expression of violent impulses. Many other explanations are, of course, possible; it is not even completely impossible that people *do* harbor many more murderous feelings toward one another than they did in 1900. But whatever the explanation of this vast change in American society (paralleled elsewhere, incidentally, and I could give other examples), genetics is not it.

Complex human conduct, then, incomprehensible without resort to meaning, is unlikely ever to be explained by reference to a single gene; at best, genetics might explain some, probably only a little, of the variation *within* a population, but not *between* different populations (and by different populations I mean different in time as well as in place, for it is as easy to be parochial about time as it is to be parochial about place). E. O. Wilson, the sociobiologist and great expert on the behavior of ants, said in his *Consilience* that he sees no reason why human history should not be reduced to laws, biological ones; indeed, he expects it to be so reduced within a relatively short time. This seems to me about as likely as Francis Fukuyama's *end* of history, which lasted but a few months. If man is a political animal, he is also a historical animal.

CHAPTER EIGHT

Yet another source of evasion, not incompatible with the neurochemical one, is provided by modern neuroscience, or should I say *neuroscientism?** New technology, which has made an enormous contribution to the practice of medicine,† is now popularly supposed to have given us a quantum leap in self-understanding as well. The results of brain scans are now regarded by much of the population with the superstitious awe that once was attached to miracle-working virgins, as providing the complete explanation of our mystery.

However, the desire for explanation always outruns the recognition of explanatory power. The illusion of understanding is more important for most men than

---

* In what follows, I am much indebted to the arguments of Professor Raymond Tallis and Dr. Sally Satel, among others.

† Not entirely beneficial, however. The habit of scanning patients until something is found, as it usually is, has become general, at least where scanners are available. Whether what is found on the scanner is the cause of the patient's complaint is a question sometimes not asked. An experienced old doctor of my acquaintance thought that the old skill of examining patients was being lost because young doctors simply put patients in what he called "the answering machine."

understanding itself; and what counts as an understanding today will be revealed tomorrow as an illusion, often so gross that people will wonder how anyone came to believe it. Conduct that was once attributed to the devil will be attributed by successive generations to a deficient or excessive phrenological bump, to an unresolved Oedipus complex, to bad conditioning, to faulty brain chemistry, to unfortunate genes, or to a part of the brain that fails to light up on fMRI scans when appropriately provoked. The illusion of understanding is like the grin of the Cheshire Cat: it is what remains when all else has disappeared.

Irrespective of the subtleties of neuroscientists, what the public understands of brain scans is that when a certain part of the brain lights up rather like traffic lights when the subject (or is it the object?) of the brain scan performs a certain task, then the part of the brain that is "responsible" for that task has been found. Thus, if I think of my mother and some part of my brain lights up (actually an averaged blood flow through that part of my brain relative to other parts, seen not at first hand but actually at nth hand), the part of my brain in which reside my thoughts and emotions about my mother is believed by many to have been found. The sheer crudity of this belief is – I was about to say astonishing, but to do so would be to disregard Mankind's long history of abject credulity in the face of authority, in this case that of miraculous-seeming machinery.

The technology is astonishing, and no one can other than marvel at its ingenuity. It promises great things, at least in the treatment of disease. But the idea that it is a direct window to men's souls is naïve, to say the least, though it is only too natural for people to suppose that when they are shown pictures of the brain with hot red patches (as nowadays they often are), they are being shown the seat, as it were, of whatever faculty of the mind has caused it to light up in this way. And if a part of the brain of a person whose behavior is abnormal in some way lights up in a scan more or less brilliantly than in persons whose behavior is normal, *voilà* – the behavior is a neurological condition, like multiple sclerosis or motor neurone disease. It wasn't me, it was my brain, will be the cry of every ill-behaved person who is minimally aware of the zeitgeist, as ill-behaved people tend to be. The fact that the same could be said of all behavior whatsoever, including that of those who seek to punish them, with as much or as little justification, will altogether escape them.

The expectations placed in neuroscience of this kind have already reached the levels of the Tulipomania, and not only among the general public. Vanderbilt University, for example, has set up a network of research into neuroscience and the law which expects scanners to be able to illuminate such matters as the capacity of a person accused of a crime – of course, a serious one, not parking where it is prohibited – to control himself and behave otherwise than he did. The assumption is that, if it can be shown that the person is neurologically deficient in self-control, he will escape punishment, since he will have been shown thereby to lack the *mens rea* necessary for a crime to have been committed at all.

Actually, the logical consequences of this approach are not as liberal as often conceived, and are precisely as illiberal as was pointed out by C. S. Lewis before scans had ever been heard of. Remove moral responsibility from the law, and what you are left with is technical administration. Does someone have a neurological defect? If so, is it changeable? If not, preventive detention is admissible. An why wait for someone with that defect to commit a crime? Why not scan everyone, and segregate those with the defect in camps, as Castro segregated those who were infected with HIV?

Of course, the unpleasant logical or practical consequences of a proposition do not make it false. A finding may have a nasty corollary but still be a true finding. However, the idea that scans will one day make all our moral distinctions for us, so that in effect we shall no longer have to think morally at all, is deeply shallow (if shallows can be deep). It amounts to a new version of the utopian vision that T. S. Eliot satirized when he said that utopians dream of a society so perfect that no one will have to be good. The neuroscientists dream of a world so neurologically healthy that no one will have to be good.

Responsibility is not just a neurological condition (though there is no doubt that brain pathology can destroy it). It is not a thing, a physical object, that can be seen, and the supposition that it is will encourage people, at least those who are ever on the lookout to excuse themselves (which is most of us), to blame not their brains, but some small part of their brains, for such of their behavior as needs excusing. (The neuroscience of saying "please" and "thank you" will interest them a good deal less.) There will never be a time when we will be able to put a person in a scanner and get a printout from it of extenuating circumstances.

CHAPTER NINE

Changes in brain scans have been found in drug addicts, from which the National Institute on Drug Abuse has drawn the conclusion that drug addiction is a chronic relapsing brain disease *and nothing else.* Actually, it had concluded this long before brain scans showed anything (since there were no brain scans when it first came to this conclusion), because only by defining addiction as a bona fide *illness,* a cross between rheumatoid arthritis and cerebral malaria, could it extract money for research from Congress. To have told the latter that addiction is only to a minor extent a physiological phenomenon, and that other contributory factors, historical, economic, cultural, legal, and social, are of much greater significance would have been to cut off the supply of that money. A purely technical solution to the problem was dangled before Congress as feasible, if only enough scientific research were done: a solution that is no nearer now than it was before billions were spent on the search.

That it is possible to show changes in the brains of drug addicts is not in the least surprising, unless one were to take the unfashionable and untenable view that brains had nothing whatever to do with behavior. The brains of London taxi drivers, for example, have been shown to differ from those of the non-taxi-driving inhabitants of London because those drivers have all mastered one of the most formidable intellectual tasks known to me (they must have done so to obtain their licenses), namely a complete topographical and cartological knowledge of the higgledy-piggledy streets of the city, such that they can immediately choose the shortest route between any two of them. However, this means that changes in the brain do not indicate that whatever causes them is a brain disease, let alone one that is chronic and relapsing, over which, by implication of those words, the person who supposedly has it can exert no control. Driving taxis is not a disease.

Heroin addicts take that drug on and off for eighteen months on average before they addict themselves to it. They thus show considerable persistence and determination in becoming addicted, which would be laudable if they were pursuing a more laudable goal. And the evidence is quite clear: notwithstanding any brain changes they may have, many established addicts give up spontaneously, largely by the old-fashioned means of taking thought. The American philosopher Herbert Fingarette pointed out in his book *Heavy Drinking* that experiments show that even very drunk chronic alcoholics (than which no addiction is more chronic) can resist the temptation to have another drink if they have an incentive to do so. Their problem is that they rarely *do* have such an incentive.

If complex behavior such as addiction is a chronic and relapsing brain disease (and it *is* complex, because it involves not just taking the addictive substance, but finding it), no one should be surprised if addicts awaited their salvation by means of a magic bullet. To imply that there *is* or *could be* such a magic bullet is, in effect, to compound the problem for addicts; for, already given to much self-deception, it is just what they want to hear so that they can continue their self-destruction with a clear conscience and that self-righteousness that comes nowadays with the awareness of being a victim – the victim of a chronic, relapsing brain disease, as revealed by brain scans.

Those who tend them, of course, also need them to be victims. This is not just a matter of financial interest: seeing victims everywhere you look is the zeitgeist, it is what gives people license to behave as they like while feeling virtuous. Virtue is not manifested in one's behavior, always so difficult and tedious to control, but in one's attitude toward victims. This view of virtue is both sentimental and unfeeling, cloying and brutal: for it implies that those who are *not* victims are not worthy of our sympathy or understanding, only of our denunciation. Thus a dialectic is set up between libertinism on the one hand and censoriousness on the other, the latter being precisely the characteristic that seeing victims everywhere, and disguising from them the degree of their own responsibility for their situation, was designed to avoid.

Be this all as it may, the type of research that shows brain changes in addicts, and concludes therefrom that they suffer from a chronic relapsing brain disease, is perfectly calculated, whether intentionally or not, to divide a man from an understanding of his own part in his own behavior. It will make him (if he takes it seriously) shallow, self-deceiving, deceitful of others, self-obsessed, and boring. It will have the natural consequence of rendering him less inclined to alter his own behavior, because it persuades him that he cannot do so. But what such a man needs is self-examination, not examination by others. No doubt people can sometimes be helped in their self-examination by means of Socratic dialogue, but such dialogue is not therapy, and indeed it is vital that it should not be professionalized, or thought of as such: for once the word *therapy* is attached to it, the whole infernal cycle is started again: "I would change, if only I could talk to Socrates."

That the latest neuroscience as a means of humanity understanding itself has been grossly oversold scarcely needs proof. It offers the illusion of understanding rather than understanding itself. No one seriously expects that human existence shortly is to be shorn of its complexities, difficulties, emotional upheavals, triumphs, and disasters as a result of the findings of neuroscience. What is reasonable to expect is that therapeutic advances for a limited number of conditions will be made, but that these advances, while reducing much suffering, will not affect the fundamental terms on which human life is conducted.

CHAPTER TEN

Of late years, another stream of thought has flowed into the great river of admirable evasions, namely that of neo-Darwinism, which claims to have succeeded where all others have failed. By placing Man's conduct in an evolutionary perspective, by subsuming it under ethology (the study of animal behavior), Darwinists believe they have plucked out the heart of our mystery. They can, of course, form alliances with neurochemists and geneticists, the latter particularly; but they believe that by returning to the great principles of natural selection and the struggle for existence, they have finally explained the previously inexplicable.

Their writings, however, are often laughably crude (though rarely laughed at): and the metaphysics of very clever men frequently makes the crudity of the uneducated seem sophistication itself. In his foreword to Richard Dawkins's book *The Selfish Gene*, Robert Trivers, one of the most important contemporary theorists of evolution, wrote:

Feeble thoughts have been strung together to produce the impression that Darwinian social theory is reactionary in its political implications. This is *very far from the truth. The genetic equality of the sexes is, for the first time, clearly established by Fisher and Hamilton. Theory and quantitative data from the social insects demonstrate that there is no inherent tendency for parents to dominate their offspring (or vice versa).*

This seems to be a variation of the Cole Porter song – the ants, bees, and wasps do it, let's do it, let's.... But do *what* exactly? What must we do in order to learn from the social insects? Build a hive? Crawl across the kitchen counter in search of sugar, as the ants do in my house in France? Mandeville published his *Fable of the Bees* in 1714, since when the political guidance provided by bees has become literal rather than metaphorical – which demonstrates how there can be intellectual regress as well as progress.* For ignorance of human life and lack of common sense it would be difficult to beat this passage.

Here, perhaps, it is worth pointing out that "the genetic equality of the sexes" referred to by Trivers is misleading, a sign of deep confusion in his mind. The sexes are not genetically identical: if they were, they would be the same sex. It is manifestly not the case that the sexes of all species are genetically determined to be equally large or equally strong. So the sexes are not genetically equal in the sense of being identical. But any other kind of equality is a moral equality, and such equality can rest only on moral arguments. Whether a biological difference is morally relevant to, say, different

---

* *Go to the ant, thou sluggard; consider her ways, and be wise* (Proverbs). Although Trivers is antireligious, he is a literalist, at least where this passage of the Bible is concerned.

political or economic treatment is itself a moral question, which no amount of further biological information can resolve.

Trivers in the same foreword made clear that the study of evolution held enormous lessons for mankind:

Darwinian social theory gives us a glimpse of an underlying symmetry and logic in social relationships which, when more fully comprehended by ourselves, should revitalize our political understanding and provide the intellectual support for a science and medicine of psychology. In the process it should also give us a deeper understanding of the many roots of our suffering.

What it will not do, however, is clear up Trivers's confusions, which are both manifest and manifold: for either human political and historical developments take place according to the laws of evolution, or they do not. If they do, then knowing the laws will make no difference, since those developments will occur anyway; and if they do not, then knowing them will be beside the point. I am reminded of the old Marxist dilemma between activism on the one hand and the ineluctability of historical processes on the other.

According to Trivers (by implication), Pope was entirely mistaken to assert that the proper study of Mankind is Man: for him, Trivers, the proper study of Mankind is earthworms and spiders. Now I am the last person to deny the fascination of all the forms of life upon earth, on which indeed the neo-Darwinists write with passion and insight; but the notion that I may decide whether to take the 9:45 bus to the station to catch the 11:45 train, and thus risk wasting an hour at the station, or the 10:45 bus, and risk missing the train altogether, by examining the lives of newts or toads (to name but two of the almost infinite variety of creatures to whom I could look to divine an answer to my dilemma) is simply preposterous. And which bus to take is only one of the simpler of my problems.

Trivers's evident belief that the theory of evolution carries within it the solution to our suffering (else why bring the matter up?) is an invitation to an infinite regress of *ad hominem* argumentation, or rather *ad bestias* argumentation, that is to say, a prolonged investigation of the biological origins of a proposition rather than an examination of the evidence in its favor: for if we could proceed directly to the latter, the former would be unnecessary and irrelevant.

Trivers was still a comparatively young man when he wrote the foreword to *The Selfish Gene*, and we all say or write foolish things when we are young. But that his development is arrested is sufficiently demonstrated by his book about the biological origins of Man's propensity to lie, *The Logic of Deceit and Self-Deception in Human Life*, published in 2011 (I take Trivers as emblematic of a whole school of thought, not for any other reason), the preface of which begins:

The time is ripe for a general theory of deceit and self-deception based on evolutionary logic, a theory that in principle applies to all species but with special force to our own.

In other words, Man is *especially* the prisoner of his biology: *Staphylococcus aureus* may have choices, but not Man. This is surely beyond the reach of satire.

Nor is this but a slip of the pen; quite the contrary, Trivers is terminally inexact in his thought. In the first few pages of his book we learn, for example, that deception "always takes the lead in life, while detection of deception plays catch-up." But his concept of deception (which, incidentally, is logically dependent upon the possibility of truth-telling) is loose and sloppy:

When I say that deception occurs at all levels in life, I mean that viruses practice it, as do bacteria, plants, insects, and a wide range of other animals. It is everywhere. Even within our genomes, deception flourishes as selfish genetic elements use deceptive molecular techniques to over-reproduce at the expense of other genes.

This passage illustrates a common feature of the neo-Darwinists: their near-panpsychism, or their attribution of purposes and goals to near-inanimate chemical objects such as viruses and genes. The one aspect of the universe for them that does not have purposes and goals – indeed is mysteriously functionless considering the length of evolutionary time necessary to have brought it about, and which is to human psychology what the appendix is to the human digestive system, that is to say, redundant – is human consciousness, for Trivers says:

neurophysiology shows that the conscious mind is more of an observer after the fact, while behavior itself is usually unconsciously initiated.

We have almost come full circle to the black box psychology of the behaviorists, though this time there is no point in paying attention to the contents of consciousness (the very thing that gives significance to human life), because it is an epiphenomenon, and is to real causes what shadowboxing is to world heavyweight championship fights. It conjures up the possibility of a man behaving in precisely the same way unconscious as conscious, not as an exception but as a rule, while all the time those clever viruses and genes *intend* to deceive – for where there is no intention, there can be no deceit.

In a fashion that is typical of the neo-Darwinists, Trivers hypostasizes an analogy and turns it into an actual, existent thing. There is no essential difference for him between HIV "changing coat proteins so often as to make mounting an enduring defense almost impossible" on the one hand and the schemes of Bernie Madoff on the other: they are both deceit. Bernie Madoff deceived his clients by a variety of ruses, including the offer (ridiculous and impossible in retrospect) of excellent, regular, but not spectacular returns on their money, which it would be their privilege to allow him to make for them, while "viruses and bacteria often actively deceive to gain entry into their hosts ... by mimicking body parts so as not to be recognized as foreign." They are both equally deceitful; but with this kind of argument, practically anything can be made to be the same as anything else. The main difference in this view between viruses and Mr. Madoff is that the viruses know what they are doing. It is they, not Mr. Madoff, who ought to be in prison.

In fact, Trivers tells us, deception and self-deception, which are the pure products of evolution, lead us to "rationalize immoral behavior." In other words, as is clear throughout his book, he has a sense of right and wrong (which does not accord with everyone else's, illustrated by his dedication of the book to the memory of Huey Newton), but this poses – or at any rate *should* pose – an insoluble dilemma for him. If evolution is a directionless, purposeless, natural phenomenon that is all-explanatory, then human behavior just *is*, it is neither moral nor immoral; if, on the other hand, it is moral or immoral, then either the process of evolution is directed by a moral intelligence in the direction of greater morality (precisely the view that neo-Darwinists wish to escape), or the process of evolution is not all-explanatory, and has led to something that escapes it, as a rocket escapes the earth's atmosphere, namely human conceptions of morality.

Evolutionists' (and neuroscientists') attempts to found morality on evolution are reminiscent of Marxist attempts to found it on the allegedly ineluctable processes of history. Like Marxists, evolutionists constantly mistake the supposed origins of a belief or conduct for its justification or lack thereof. They are rather someone who would put a person in a scanner to establish whether, when he says that two and two make four, two and two actually *do* make four. As Doctor Johnson said, argument is to be invalidated only by argument.

Despite the great intellectual brilliance of the neo-Darwinists, their ideas, at least about human life, are often at base of an astonishing crudity. They write like people who know that humans exist, but have never actually made contact with any. Trivers, for example, says:

In this book we take an evolutionary approach to the topic. What is the biological advantage to the practitioner of self-deception, where advantage is measured as positive effects on *survival and reproduction? How does self-deception help us survive and reproduce – or, slightly more accurately, how does it help our genes survive and reproduce?*

Now according to the neo-Darwinians, this is a question that must be asked of all human behavior whatsoever, natural selection being, according to Daniel Dennett, the intellectual philosopher's stone that unites "the realm of life, meaning and purpose with the realm of space and time, cause and effect, mechanism and physical law." In essence, the only question to be asked of anything relative to human existence is: how does it help the genes survive in the largest number? Mr. Madoff's schemes, then, were ultimately to procure numerical advantage of the genes that, temporarily, took a ride in him, but also caused him to behave as he did – the genes of the detectives who caught him behaving in precisely the same way. It is no good enquiring whether Mr. Madoff actually, as a matter of empirical fact, had more children than the detectives who helped put an end to any possibility of any further reproductive career on his part, because of course what we must estimate is gene survival, not the survival of the organismal envelopes known to naïve and biologically uninformed people as – well, people. And the genes that caused Mr. Madoff to swindle untold numbers of people may have done magnificently, notwithstanding his long jail sentence, while those of the poor detectives are doomed to extinction.

This is all so preposterous that it hardly needs, or merits, refutation. But we can easily see how this kind of nonsense, with its smokescreen of elaboration and

sophistication, serves to lessen the pains of Man's responsibility for himself both individually and collectively, and
is yet another admirable evasion of whoremaster man,
laying his goatish disposition to the charge of a star.

And that is what psychology almost inevitably leads
to: evasion after evasion.

CHAPTER ELEVEN

I do not mean to argue, of course, that psychological or pharmaceutical therapies never assist anyone, though the theories upon which they are based may be entirely mistaken (witch doctors or exorcists may effect cures, but that does not mean that the spirits, ghosts, and devils that they claim to have driven out actually existed). Used sparingly and with discretion, they may be very useful to carefully selected individuals. Unfortunately, such is the self-aggrandizing nature of most modern "caring" professions that alleged competence in and sovereignty over matters which are beyond the reach of technical understanding or solution undermine any residual modesty, realism, or judgment that they might otherwise still have had. No practitioner now says, as does the doctor in *Macbeth*:

> Therein the patient
> Must minister to himself.

And therefore no patient says in return:

> Throw physic to the dogs, I'll none of it...

but rather continues on his help-seeking way no matter how many times real help fails to be forthcoming. At the very moment I wrote this I received via the Internet the *New York Times* for the day, which contained an article about violent and disruptive boys. The article described the plight of the parents, indeed a terrible one:

No single diagnosis fits, no drug brings real relief, and if the teenager rejects the very idea of psychotherapy, there is little chance of lasting improvement.

But yet they never lose faith in the possibility of a technical fix, and indeed still demand it as a right.

I do not wish to deny that psychologists have done many intriguing and ingenious experiments. But the overall effect of psychological thought on human culture and society, I contend, has been overwhelmingly negative because it gives the false impression of greatly increased human self-understanding where none has been achieved, it encourages the evasion of responsibility by turning subjects into objects where it supposedly takes account of or interests itself in subjective experience, and it makes shallow the human character because it discourages genuine self-examination and self-knowledge. It is ultimately sentimental and promotes the grossest self-pity, for it makes everyone (apart from scapegoats) victims of their own behavior, precisely as Edmund in *Lear* says.

Psychology devotes itself to the search for its holy grail, the new spherical predominances, planetary influences, and divine thrustings-on that supposedly explain ourselves to ourselves, them to us, and us to them.

Let me take some of the most fascinating psychological experiments known to me, those of Stanley Milgram, published in his justly famous book, *Obedience to Authority*. These experiments asked ordinary members of the public to administer, at the experimenter's behest, electric shocks to a human experimental subject supposedly engaged upon a learning task. When the subject made an error, the member of the public applied what he supposed was an electric shock to him, of increasing voltage with each successive mistake. Milgram discovered that the majority of ordinary, God-fearing folk would administer shocks so severe that, if they had been real rather than simulated, they would have killed the experimental subject. The book can be read in almost the same way as a sophisticated thriller.

As it happens, I was commissioned by a magazine to write an article about the book on the thirtieth anniversary of its publication. I took it with me on a flight from England to Ireland, and sat next to an Irish social worker. She saw the title of the book, which she had not read, and said to me that she had always been against all forms of authority because she had grown up in Ireland when it was still heavily under the social, political, and economic control of the Catholic Church.

"You are against all authority?" I repeated.

"Yes," she replied.

"So you don't mind if I now enter the cockpit and take over the piloting of this plane?"

"I don't mean that kind of authority."

What she had meant, of course, was that she was against the kind of authority that is excessive or oppressive. We do not really need experiments to show us that excessive authority is harmful, and only a moment's reflection is necessary to demonstrate the need for authority nonetheless. Getting the balance right – *right* being an irreducibly moral category, incidentally, not to be found experimentally or on the red patches of a scanned brain – is a difficult matter. Milgram's experimental findings would not horrify us, nor even interest us, were we not convinced that shocking strangers to death is wrong.

Milgram's experiments surprised people because of the unexpectedly large number of people who were apparently willing to administer dangerous and even potentially fatal electric shocks to strangers on the mere orders of an experimenter in a white coat. But the lessons to be drawn from this and their applicability to real life beyond the experimental situation are not at all clear. That authority can often be used for ill purposes is the common experience of mankind, as is, perhaps slightly less frequently, its opposite, that it is often used for good. It can switch from one to the other, moreover: what starts out as legitimate or good purpose can become illegitimate or bad purpose. Some taxation is justified, but not all is. No sifting of psychology will by itself illuminate which is which.*

The more we examine Milgram's experiment, the less certain are we that we can derive anything from it that we could not have derived from elsewhere, in fairly peremptory manner at that, merely by considering a little of history or of literature. And with the exception of

---

* The fact that I do not myself have any watertight metaphysic of morals does not mean that psychology can just rush in to fill the gap.

specific instances, psychology has contributed nothing to human self-understanding; rather the reverse: for by coming between a man and what Doctor Johnson called "the motions of his own mind," it acts as an obstacle to genuine (though often painful) self-examination.

Literature does not do this. Of course, not all literature is an aid to self-examination, and some of it has promoted foolishness or rank bad ideas; but even this is generally done by suggestion rather than by direct inculcation of a fatuous doctrine.

I shall give but two examples from literature that assist us to a deeper understanding of our own existence and that of the people around us. I use them not because they are necessarily the best possible examples, but because they come most immediately to my mind. The first is a passage from *King Lear* that I have already quoted. Lear in his madness says:

> Thou rascal beadle, hold thy bloody hand.
> Why dost thou lash that whore? Strip thine own back.
> Thou hotly lust'st to use her in that kind
> For which thou whipp'st her.

These words, written four hundred years ago, illustrate the way in which we may – not *must* – project on to others our own failings or wickedness, attributing to them our own illicit desires or secret thoughts. Having vehemently denied that those desires are our own, we then have what the Freudians call a reaction formation, an exaggerated hatred of the object of our desires, that can on occasion lead to cruelty through excess of zeal. And surely all are familiar with vehement denial of what we secretly know to be true. The difficulty is that vehemence is not *necessarily* a sign of such denial, though it can be, and often is. Honest examination of ourselves, and shrewd (though not infallible) appraisal of others, is what is necessary to distinguish between the two possibilities. Nothing in this respect has changed since Shakespeare's day; the poet gives us understanding without providing us with an excuse for our own conduct; rather the reverse. Four lines of Shakespeare are worth a bookful of Trivers.

My second example is from *Rasselas*, Doctor Johnson's short philosophical novella that, despite being written in only a few days in order to pay for his ill mother's medical treatment (in the event it paid for her funeral), has lost nothing of its wisdom in the quarter millennium since it was published. That is because, as Francis Bacon said, "reading makes a full man, conversation a ready man, and writing an exact man," and Johnson was all three.

Rasselas, Prince of Abyssinia, travels the world in search of a way of life that is perfect. Each place that he visits seems to him at first to be full of promise, but on closer examination he finds that every place, and every condition of Man, has its drawbacks, disadvantages, discontents, and inconsistencies. In Cairo, Rasselas finds a professor of philosophy who eloquently preaches a noble stoicism.

He showed, with great strength of sentiment, and variety of illustration, that human nature is degraded and debased when the lower faculties predominate over the higher; that when fancy, the parent of passion, usurps the dominion of the mind, nothing ensues but the natural effect of unlawful government, perturbation and confusion; that she betrays the fortresses of the intellect to rebels, and excites her children to sedition against reason, their lawful sovereign. He compared reason to the sun, of which the light is constant, uniform and lasting; and fancy to a meteor, of bright but transitory lustre, irregular in its motion, and delusive in its direction.

Rasselas is deeply impressed:

[He] listened to him with the veneration due to the instructions of a superior being...

and he tells his guide during his peregrinations, Imlac, that:

"I have found ... a man that can teach all that is necessary to be known..."

Imlac warns him to "Be not too hasty ... to trust, or to admire, the teachers of morality: they discourse like angels, but they live like men."

But Rasselas, being young, does not heed the warning because he "could not conceive how any man could reason so forcibly without feeling the cogency of his own arguments..."

The day following the lecture, Rasselas visits the philosopher at his home and finds him utterly disconsolate because "my daughter, my only daughter, from whose tenderness I expected all the comforts of my age, died last night of a fever." He continues, "My views, my purposes, my hopes are at an end..."

Rasselas, still full of the lecture, replies in best callow-youth fashion:

"Sir ... mortality is an event by which a wise man can never be surprised: we know that death is always near, and it should therefore always be expected."

To this the philosopher returns a cri de cœur which the American Psychiatric Association would do well to note:

"Young man ... you speak like one that has never felt the pangs of separation."

Rasselas persists a little in the rational stoicism that he has learned at the philosopher's feet:

"Have you, then, forgot the precepts ... which you so powerfully enforced? Has wisdom no strength to arm the heart against calamity? Consider that external things are naturally variable, but truth and reason are always the same."

The philosopher replies:

"What comfort ... can truth and reason afford me? of what effect are they now, but to tell me that my daughter will not be restored?"

Rasselas, reproved, "went away convinced of the emptiness of rhetorical sound, and the inefficacy of polished periods and studied sentences."

What Johnson captures so brilliantly is the inherent tragic dimension of human existence, a dimension that only literature (and other forms of art), but not psychology, can capture, and which indeed it is psychology's vocation to deny and hide from view with a thin veneer of science. Without an appreciation of the tragic dimension, all is shallowness; and those without it are destined for a life that is nasty and brutish, if not necessarily short.

INDEX

A NOTE ON THE TYPE

ADMIRABLE EVASIONS has been set in Quadraat Sans, an extension of Fred Smeijers's 1992 Quadraat family. By combining the elegance of Renaissance models with features from more contemporary types like Times and Plantin, Smeijers sought to create a family of pleasing, readable types suited to a variety of uses. First released in 1996 and extensively expanded and improved since then, the sans serif branch of the Quadraat family is designed to work in concert with the serif version and as an independent family. Both serif and sans serif faces combine excellent readability characteristics with elegant drawing, economical character fitting, and broad language support, making them ideal types for print projects ranging from books and periodicals to advertising and identity materials. ◊ The display type is DIN Next Slab, a serif adaptation of the widely popular DIN sans serif types, designed by Akira Kobayashi, Tom Grace, and Sandra Winter.

DESIGN & COMPOSITION BY CARL W. SCARBROUGH

Printed in the USA
CPSIA information can be obtained
at www.ICGtesting.com
JSHW012040140824
68134JS00033B/3170